Probability

Methods and measurement

OTHER STATISTICS TEXTS FROM CHAPMAN AND HALL

Further information of the complete range of Chapman and Hall statistics books is available from the publishers.

Probability

Methods and measurement

Anthony O'Hagan

University of Warwick

London   New York

CHAPMAN AND HALL

First published in 1988 by Chapman and Hall Ltd
11 New Fetter Lane, London EC4P 4EE
Published in the USA by Chapman and Hall
29 West 35th Street, New York NY 10001

© 1988 A. O'Hagan

Printed in Great Britain at the
University Press, Cambridge

ISBN 0 412 29530 X (hardback)
    0 412 29540 7 (paperback)

British Library Cataloguing in Publication Data

O'Hagan, Anthony
    Probability: methods and measurement.
    1. Probabilities
    I. Title
    519.2   QA273

    ISBN 0 412 29530 X
    ISBN 0 412 29540 7 Pbk

Library of Congress Cataloging in Publication Data

O'Hagan, Anthony.
    Probability: methods and measurement.
    Bibliography: p.
    Includes index.
    1. Probabilities.   I. Title.
QA273.034   1988   519.2   87-29999
ISBN 0 412 29530 X
ISBN 0 412 29540 7 (pbk.)

To Anne

Contents

Preface

This book is an elementary and practical introduction to probability theory. It differs from other introductory texts in two important respects. First, the personal (or subjective) view of probability is adopted throughout. Second, emphasis is placed on how values are assigned to probabilities in practice, i.e. the measurement of probabilities.

The personal approach to probability is in many ways more natural than other current formulations, and can also provide a broader view of the subject. It thus has a unifying effect. It has also assumed great importance recently because of the growth of Bayesian Statistics. Personal probability is essential for modern Bayesian methods, and it can be difficult for students who have learnt a different view of probability to adapt to Bayesian thinking. This book has been produced in response to that difficulty, to present a thorough introduction to probability from scratch, and entirely in the personal framework.

The practical question of assigning probability values in real problems is largely ignored in the traditional way of teaching probability theory. Either the problems are abstract, so that no probabilities are required, or they deal with artificial contexts like dice and coin tossing in which probabilities are implied by symmetry, or else the student is told explicitly to assume certain values. In this book, the reader is invited from the beginning to attempt to measure probabilities in practical contexts. The development of probability theory becomes a process of devising better tools to measure probabilities in practice. It is then natural and useful to distinguish between a true value of a probability and a practical measurement of it. They will differ because of the inadequacy or inaccuracy of the measurement technique. Although natural, this distinction between true and measured probabilities is not found in other texts. I should emphasize that the notion of a true probability is not intended to be precisely or rigorously defined, in the way that philosophers of science would demand. It is merely a pragmatic device, in the same way as any other 'true measurement'. For instance, attempts to define length to absolute precision will run into difficulties at the atomic level: atoms, and the particles within them, do not stay in rigid positions. Yet in practice we are undeterred from referring to 'the' length of something.

I have assumed only a very basic knowledge of mathematics. In particular, for the first six chapters no calculus is used, and the reader will require only some proficiency in basic algebra. Calculus is necessary for Chapter 7, and to some extent for the last two chapters, but again only an elementary knowledge is required.

Two devices are used throughout the text to highlight important ideas and to identify less vital matters.

Displayed material
All the important definitions, theorems and other conclusions reached in this book are displayed between horizontal lines in this way.

{ ....

Asides
This indented format, begun and ended with brackets and trailing dots, is used for all proofs of theorems, many examples, and other material which is separate from the main flow of ideas in the text.

...........}

There are over 150 exercises, almost every one of which is given a fully worked solution in the Appendix. You should attempt as many of the exercises as possible. Do not consult the solutions without first tackling the questions! You will learn much more by attempting exercises than by simply following worked examples.

Many people have generously given me help and encouragement in preparing this book. To thank them all would be impossible. I must mention Dennis Lindley and Peter Freeman, whose critical reading was greatly appreciated. Nor must I forget my family and their long-suffering support. I am sincerely grateful to all these people.

University of Warwick, Coventry
August 1987

1
Probability and its laws

The purpose of this chapter is to define probability and to derive its most fundamental laws. We shall do this with a minimum of fuss, in order to start the reader on evaluating real probabilities. In Section 1.1 we discuss what probability represents, namely a measure of the degree of belief held by a given person in the truth of a proposition. We also establish the probability scale as being from zero to one. A proposition has probability one if it is believed to be certainly true, and probability zero if it is believed to be certainly false.

In Section 1.2 we present the simplest and crudest technique for measuring probabilities. (More refined methods are developed later.) This technique will be referred to as direct measurement. Sections 1.3 to 1.6 are devoted to a more formal presentation of probability through betting behaviour. Section 1.3 discusses betting generally, including the many pitfalls of basing a definition of probability on gambles. A notion of a fair bet is introduced in Section 1.4 and used to define probability. This definition is then employed to obtain the two fundamental laws of probability, in Sections 1.5 and 1.6. Section 1.7 introduces the important concept of independence.

1.1 Uncertainty and probability

Uncertainty is a familiar, everyday phenomenon. As I write these words, I can quickly call to mind a great many things about which I am uncertain. Here are some examples:

The weight of the British Prime Minister.
Who will be the next President of the United States of America.
Whether Newton really was hit on the head by an apple.
How many countries possess nuclear weapons.
Whether bank interest rates will fall next month.
Whether it rained in Moscow yesterday.
The largest prime number less than 10^{10}.
Whether cigarette smoking causes lung cancer.

All these questions have answers, and no doubt there are people who know the answers to some of them. But at this moment I am not certain about any of them. If I were asked to express opinions, I would avoid giving the impression of certainty. This is easy in everyday speech, because there are many forms of expression that convey appropriate levels of uncertainty. For instance:

'It is unlikely that Newton really was hit by an apple.'

'Probably at least six countries have nuclear weapons.'

'It is almost certain that smoking causes lung cancer.'

Such expressions are acceptable in ordinary conversation, because accuracy is not important. 'Unlikely', 'probably' and 'almost certain' are not intended to have precise meanings. However, there are occasions when it is necessary to measure just how likely it is that a given proposition is true.

For instance, a matter of great public interest at present is the safety of nuclear power stations. Many people have devoted considerable ingenuity to answering questions such as 'Is a Pressurized Water Reactor more likely to suffer a serious release of radioactive materials than an Advanced Gas-cooled Reactor?'

Probability is a measure of uncertainty, and the theory of probability, with the related subject of statistics, provides a scientific methodology for evaluating probabilities. Its importance can scarcely be overstated. The abundance of things about which we are uncertain, and the way in which our future actions, prosperity and even existence depend on uncertainties, is reinforced by every news report. Accurate weighing of uncertainties is vital to good decision-making at every level – personal, national or international.

How should our measure of uncertainty behave? Consider the proposition that it rained in Moscow yesterday. Because different people will have different information, they will have different degrees of belief in the truth of the proposition. One person, a Muskovite, perhaps, who got wet yesterday, may know that it did rain in Moscow. His degree of belief is the highest possible – he is certain that the proposition is true. Others may be more or less uncertain. I know almost nothing about the weather in Eastern Europe yesterday, and my belief in the proposition is much lower. A third person may remember seeing a weather map showing an area of high pressure over the Ukraine; he might consequently have very little belief in this proposition. In fact, he might believe it almost certain to be false.

Given any proposition, and at any time, a person's belief in its truth may range from feeling certain that it is false (the lowest possible degree of belief), through stages commonly expressed by terms like 'unlikely', 'probable', 'almost certain', to feeling certain that it is true (the highest possible degree of belief). We are going to replace these verbal descriptions of belief by numbers called *probabilities*. The lowest number will represent the lowest degree of belief and the highest number will represent the greatest possible belief. We could use any range of numbers for our probability scale, but scientists have found that the most convenient scale is from zero to one. Therefore a proposition will be assigned a probability of zero if it is certain to be false, and a probability of one if it is certain to be true. Probabilities between zero and one will correspond to all the various degrees of belief between these two extremes.

The probability scale
Probabilities are numbers between zero and one. Probability zero represents a proposition which is certain to be false. Probability one represents a proposition which is certain to be true. Between these limits, higher probability represents a proposition which is more likely to be true.

Several other important facts are also implicit in things we have said so far. One is that probability is a degree of belief in some *proposition*. A proposition is simply any kind of statement which may be either true or false. Traditionally, probability has been concerned with scientific applications where interest lies in the probabilities that an experiment will yield specific outcomes. Consequently, most books define probability as relating not to propositions, which are either true or false, but to events, which either occur or do not occur. The difference in terminology is not important, but our treatment of probability will be different from traditional ones, and there are various reasons for preferring 'proposition' to 'event'. One is that although scientific applications are still our main interest, we shall want to emphasize that the scope of probability is much wider. The notion of a proposition is more general than the traditional notion of an event, and therefore suits our purpose well.

Another vital fact is that probability is *personal*: different people have different degrees of belief in, and hence different probabilities for, the same proposition. We have already remarked on this in the case of the proposition that it rained in Moscow yesterday, and it is obviously true generally. The reason is that people have different information on which to base their beliefs. Furthermore, a person's information changes through time, and as he acquires new information, so his beliefs change.

Interpretation of probability
A probability is a numerical measure of degree of belief in the truth of a *proposition*, based on the *information* held by a person at some time.

This interpretation finds expression in our notation. The probability of a proposition E for a person having information H will be denoted by $P(E|H)$. The vertical bar, $|$, is read as 'given', so that $P(E|H)$ would be read as 'the probability of E given H'. Probability is personal because two persons' information can never be identical.

Notation for probability
The probability of a proposition E based on information H is denoted by $P(E|H)$, read 'the probability of E given H'.

We shall use symbols like E, F, G to denote propositions. This agrees with traditional treatments, where E stands for an event. H comprises all of the person's experiences, going right back to his childhood – the letter H can be thought of as standing for 'history'. So, H can only ever be defined implicitly by reference to the total information of a given person at a given time. However, it can be modified by the inclusion or exclusion of some specific information. For example, one may ask what degree of belief a person would have about a certain proposition if, in addition to his current knowledge, he were to learn some specified new information.

Whenever we want to discuss probabilities, throughout this book, it will be necessary to specify H as information held by some person. Yet because this book is about general theory, much of what we have to say about probabilities applies regardless of whose probabilities they are. Therefore we introduce a kind of abstract, impersonal person called You. The capital letter serves to distinguish You, the person whose probabilities we are considering, from you, the reader. The use of You emphasizes the personal nature of probability but allows us to study probability theory in a formal and abstract sense. Unless otherwise stated, the information H will always be Your information.

1.2 Direct measurement

So far we have said very little about the probability scale. We have stated its end-points, zero and one, and loosely defined their meaning, which also determines the direction of the scale. We have not yet given a proper definition of probability, and in particular we have not said what degrees of belief the various numbers between zero and one should represent. One definition is presented in this section, as a means of practical probability measurement. In Section 1.4 a second definition is given which is more useful for developing the laws of probability. The matter of the equivalence of the two definitions is left to Chapter 2.

Consider an everyday example of measurement – weighing an object by means of a traditional balance. An arm is pivoted in its centre and carries a pan at either end. Into one pan we place the object to be measured, and into the other we place an object whose weight is known. Depending on which way the balance arm tips, we can determine whether the weight of the object is greater or less than the fixed weight. Provided we have enough objects of different fixed weights, we can determine the weight of any other object to any desired accuracy. In fact, this weighing procedure could be taken as a definition of the weight of the object.

We shall call this method of measurement, in which an object is measured by comparing it with a set of reference objects whose measurements are assumed known, *direct measurement*. Measuring length by using a ruler or tape measure is another example of direct measurement. The marks on the ruler or tape measure determine the set of fixed lengths, and putting it alongside the object to be measured permits comparison with *all* the known lengths simultaneously. This property is so useful that numerous other measurements are commonly converted to lengths, so that they can be measured by reference to a scale. The thermometer is an example. It is also an example of what we shall call *indirect measurement*. The quantity to be measured influences some other quantity (typically a length) which is then measured, and the original measurement is deduced from some known relationship between the two. Our second definition of probability in Section 1.4 is in terms of an indirect measurement, whereas in this section we consider direct measurement of probabilities.

For direct measurement we shall need a set of known measures. In the case of probability we require a set of reference propositions, i.e. propositions whose probabilities are assumed already known. Suppose that You have before You various bags, each labelled with two numbers. One number is written in red and the other in white. The bags contain various quantities of red and white balls, as given by the red and white numbers. For instance, if a bag is labelled with a red 3 and a white 2 then that bag contains 3 red balls and 2 white balls. A ball is to be taken out of this bag. Let $R(3,2)$ denote the proposition that the chosen ball is red. Or in general, let $R(r,w)$ denote the proposition that a red ball is drawn from the bag labelled with the number r in red and the number w in white (i.e. the bag containing r red balls and w white balls). You may assume (as part of Your information H)

(a) that the balls are identical in every respect except colour,
(b) that the bag will be shaken vigorously to disturb all the balls before the chosen ball is drawn out, and
(c) that the draw will be made blindfold.

Reference probabilities

$$P(R(r,w)|H) = \frac{r}{r+w}.$$

In other words, the probability of drawing a red ball from a bag equals the proportion of red balls in that bag.

The propositions $R(0,1)$, $R(1,0)$, $R(1,1)$, $R(2,1)$, $R(1,2)$, and so on, constitute our set of reference propositions. You have bags for all combinations of r and

w, so that the reference probabilities $r/(r+w)$ will cover all the rational numbers between 0 and 1 inclusive. The reference probabilities agree with what we have said so far about probabilities. The probability zero arises when $r/(r+w)=0$, i.e. when $r=0$, and it is clearly impossible to draw a red ball from a bag containing only white balls. Similarly, probability one corresponds to drawing a red ball from a bag containing only red balls ($w=0$). Between these two extremes, You will have a greater degree of belief in (and therefore a greater probability for) drawing a red ball the greater the proportion of red balls that are in the bag.

Having defined our set of reference propositions, we must now consider how they are to be compared with any required proposition E in order to determine Your probability $P(E|H)$. By definition, $P(E|H)$ is greater than $r/(r+w)$ if, and only if, Your belief in the truth of the proposition E is greater than in the truth of $R(r,w)$. Unfortunately, Your degrees of belief are inside Your head, and so the comparison is not easy. Nevertheless, for the present we will ask You to compare E with $R(r,w)$ simply by thinking about them. You must ask Yourself which You consider more likely, that E is true or that a red ball will be drawn from a bag containing r red and w white balls.

Direct measurement definition of probability

To measure Your probability $P(E|H)$ You compare E with the reference propositions $R(r,w)$. For every proposition $R(r,w)$ for which You have a lower degree of belief than in the truth of E, $P(E|H)>r/(r+w)$.

Let us admit straight away that direct measurement of probabilities is usually very unsatisfactory. It is like comparing the weight of an object with a reference weight by placing one object in each outstretched hand and trying to say which one *feels* the heavier. Our defence is simply that we are starting with the crudest of all devices for measuring probabilities. More sophisticated devices will follow, and to construct them is the main intention of this book.

To illustrate direct measurement, consider a simple example. Suppose that I wish to determine my probability for a 'white Christmas'. Accordingly I let E be the proposition of snow falling in a given place, say central London, on Christmas Day this year. H will represent all my information as usual. To determine $P(E|H)$ I first compare E with $R(1,1)$. I decide that in my opinion E is less likely than drawing the red ball from a bag containing one red and one white ball. Therefore $P(E|H)<0.5$. In fact, I think E is less likely than drawing the red ball from a bag containing one red and nine whites, so $P(E|H)<0.1$. But comparing E with $R(1,19)$, I decide that E is more likely, and so $P(E|H)>0.05$. I might next consider $R(1,14)$, to see whether $P(E|H)$ is greater or less than one-fifteenth, and by continuing in this fashion I can (at least in principle) determine $P(E|H)$ to any desired accuracy.

Practice using direct measurement of probabilities. Although we shall construct more elaborate measurement techniques later, almost always *some* probabilities must be measured directly. So any practice you obtain in direct measurement will be valuable. For instance, every week during the winter months a large number of football matches will be played in almost every country in Europe and South America, and in many other countries also. For each match, measure your probability for each of the three propositions W: that the home team will win, L: that the away team will win, and D: that the match will be drawn. Devotees of other sports may adapt these three propositions to their own sports. Alternatively, you may be interested in the weekly charts of popular music. Each week, for each record currently in the top twenty, consider the three propositions R: that the record will rise higher next week, F: that it will fall lower, and S: that it will stay in the same position. Making regular probability evaluations for triples like (W, L, D) or (R, F, S) is very good practice.

You should also remember that probability applies wherever there is uncertainty. Make a habit of measuring probabilities for the countless uncertainties of everyday life. If you go to a bookcase to get a book, before looking ask yourself: 'What is my probability that the book is on the top shelf?'. When you go to buy a pair of shoes: 'What is my probability that I will buy a pair at the first shop I enter?'. When you answer the telephone: 'What is my probability that this is a call from my mother?'.

Do not, of course, try to measure these probabilities at all accurately. In principle, direct measurement can be arbitrarily precise, which is necessary for it to serve as a definition. In practice it is quite different! Most people will have considerable difficulty in making comparisons. In the case of the proposition E of a 'white Christmas', having decided that $0.05 < P(E|H) < 0.1$, I would find it impossible to go much further than this. I simply cannot say definitely whether my degree of belief in E is stronger or weaker than my degree of belief in $R(1,13)$. The best I can say is that my probability $P(E|H)$ is 'approximately $1/14$' or 'about 0.07'. Or, to within the level of accuracy that I can achieve with direct measurement, $P(E|H) = 0.07$.

You need not always use the 'balls in the bag' reference probabilities for comparisons. Having evaluated your probability for one proposition, you can then use it as a comparison when evaluating other probabilities. For instance, suppose that you decide to give a probability of 0.4 to the proposition of team A beating team B. If you feel that team C is more likely to beat team B than team A is, then you will give a probability greater than 0.4 to this proposition. It is easier to compare propositions of similar type than different types, so this kind of direct measurement is particularly useful when you are evaluating probabilities for a series of related propositions.

Exercises 1(a)

1. Consider the proposition E, that Shakespeare wrote more than 50 plays. As a first step in measuring your probability for E compare E with the reference proposition $R(1,1)$ and thereby determine whether your probability is greater or less than one half.

2. Continue your measurement of your probability for E, comparing it with $R(r,10-r)$ for $r=1, 2, \ldots, 9$, so as to obtain a measurement to within 0.1.

3. Alter your information H by finding out more about Shakespeare's plays. List all the plays that you know, ask your friends, go to a bookshop or library and find more titles. Then measure your new probability as in 1(a)2.

4. Take an ordinary coin and imagine that you are about to toss it a number of times. Measure directly, as accurately as you think is reasonable, your probabilities for the following propositions.
  C_1: that the first toss will fall Heads.
  C_2: that the first three tosses will all fall Heads.
  C_3: that in the first five tosses at least three will fall Heads.
  C_4: that in the first six tosses exactly three will fall Heads.

5. Toss your coin six times and see whether proposition C_4 is true. Measure your probability for the proposition C_4', that in the next six tosses exactly three will be Heads.

6. Toss your coin sixty more times, and in each successive set of six tosses see whether three turn out as Heads. Now measure your probability for C_4'', that in a further six tosses three will be Heads.

1.3 Betting behaviour

For the remainder of this chapter we give our attention to a more formal definition of probability. We shall use an indirect measurement, defining probability in terms of betting behaviour. The idea is, that Your degree of belief in a proposition causes You to accept or reject bets which are offered to You. A measure of willingness to bet is then converted into a measure of degree of belief, i.e. probability. A bet has three constituents,

  1. the *proposition*, which determines the outcome of the bet,
  2. the *odds* which are offered by the bookmaker, and
  3. the amount of money You are prepared to *stake*.

The terms 'odds' and 'stake' are defined simply as follows.

Betting terms

A bet between You and a bookmaker on a proposition E at odds z and with stake $£s$ means that if E turns out to be false You will give the bookmaker $£s$, and if E turns out to be true the bookmaker will give You $£zs$.

The stake is the amount of money which You stand to lose if the proposition is not true. The odds represent the ratio of the amount You stand to win if the proposition is true divided by the amount You stand to lose. In everyday betting, the odds are expressed as 'x to y' where x and y are two integers and our number z is the ratio x/y. The reason for this cumbersome usage is historical; people making frequent bets did not usually understand fractions. The device is unnecessary for a mathematically literate audience and so, for us, odds will be a single, non-negative number z.

We are going to measure Your probability $P(E|H)$ by means of bets on E between You and a hypothetical bookmaker. Suppose that You are offered a selection of bets on E at various odds and a fixed stake, say $s=1$. You must say which bets You accept and which You refuse. Now since the stake is fixed, the amount You stand to lose is £1 in each bet. What varies is the amount, $£z$, that You win if E is true. Obviously, You will accept any bet for which z is sufficiently large, and reject all others. The expression 'sufficiently large' is rather vague, but is adequate to bring out an important fact, that what You consider sufficiently large depends upon Your degree of belief in the truth of E. If You believe that E is almost certainly true then You do not expect to lose Your £1 stake. Then even a bet at low odds, offering only a small reward, may be acceptable. If, however, You believe that E is almost certainly false then You anticipate losing Your stake. It would need a high odds value to induce You to accept the offered bet.

This discussion suggests that we try to define formally the notion of 'sufficiently large' odds, in order to measure degrees of belief. However, there are a number of practical difficulties with this approach. First, we have ignored the matter of stake. What You regard as sufficiently large odds depends on Your stake. If the stake is high then You are being offered a bet with serious financial consequences to You if it turns out that E is false. In general, the higher the stake the higher must be the odds before they are 'sufficiently large' to be acceptable. Clearly, imposing a high stake distorts Your betting behaviour, and if we tried to measure probability through betting behaviour then our measure would also be distorted. It seems sensible, then, to work only with low, financially insignificant, stakes. But what is a low stake for one person may be a high stake for a much poorer person. In order to ensure that the stake is low enough, we have to imagine a stake of perhaps one penny rather than one pound, or one cent rather than one dollar. Then we hit different problems. When the stake is too low, You stand to lose so little that it hardly matters what bets You accept.

The phenomenon of higher stakes demanding higher odds is related to an even more damaging practical difficulty. It is known that almost everybody is to some extent 'risk averse'. This means that not only will You reject bets that You regard as having unfavourable odds, You will also reject bets at apparently favourable odds unless they are sufficiently favourable to compensate You for the risk of losing Your stake. Higher stakes require a higher degree of compensation in terms of favourable odds. The importance of risk aversion can be demonstrated by a simple example. Insurance companies make profits. They do so because the premiums they charge are higher than the costs necessary on average to cover the insured risks. People taking out insurance are prepared to pay the extra money simply to avoid the risk of a much greater financial loss. The profit compensates the insurance company for taking on the risk. Because the company is much richer than the individual, it is less risk averse, and so the amount the individual is prepared to pay to avoid the risk is more than enough to compensate the company for accepting it. Both parties are happy. The same game, played at all levels from individuals to nations, ensures that the rich continue to get richer and the poor poorer.

Because of risk aversion Your 'sufficiently large' odds will reflect Your degree of belief only in a rather complicated way. It is certainly possible to build a theory of probability on actual betting behaviour, but an extra concept is needed. By introducing *utilities* we can explain risk aversion and the fact that some people are more risk averse than others, and a great deal more besides (including compulsive gambling). An excellent elementary account of utility is given by D. V. Lindley, *Making Decisions* (John Wiley, 1985). A rather more advanced treatment which presents probability and utility in a unified development is M. H. deGroot, *Optimal Statistical Decisions* (McGraw-Hill, 1970).

We take a different route, avoiding the added complication of utility. We shall define probability by reference to a kind of idealized betting behaviour.

1.4 Fair bets

Consider a particular, very familiar proposition. A coin is to be tossed in the usual way; let E be the proposition that the result is Heads. You will be offered bets on E with a fixed stake (say £1, or whatever seems a moderate amount to You). Consider odds of 0.5, i.e. Your potential gain is only half of Your stake. This is clearly unacceptable (unless You have a superstitious faith in Heads). On the other hand, for most people odds of 2 would be very acceptable. Therefore, for most people a sufficiently large odds value lies between 0.5 and 2. Odds of 0.5 is clearly unfavourable to You, whereas odds of 2 is favourable to You, and therefore unfavourable to the bookmaker. Now instead of asking which bets You would actually accept, we ask only what odds You regard as favourable to You. This avoids questions of risk aversion and stake size. In the present example the minimum odds at which people would actually bet on a coin toss will depend on the required stake and vary from person to person. It could easily range from 1.1 to 1.5 or beyond. Despite this, almost everyone would accept, because of the symmetry of the situation, that odds of 1.0 is actually fair. Odds greater than one favours You, and odds less than one favours the bookmaker.

We base our definition of probability on this notion of *fair odds*. In the last paragraph we chose to examine a special situation in which fairness is very easy to judge, and therefore we could expect agreement between most people. In other cases, what one person judges to be fair odds for a proposition another may regard as highly favourable to one or other party, reflecting the fact that people's degrees of belief in a proposition vary. Our notation, as for probabilities refers to the knowledge or information H.

Odds notation

Your fair odds for a proposition E based on information H is denoted by $O(E|H)$.

Having defined fair odds as a proxy, or indirect, measurement for degrees of belief, we need a rule for transforming it to probabilities. The general form that this rule should take is easily deduced from characteristics of the odds scale. The lowest possible value for $O(E|H)$ is zero, and corresponds to E being considered certainly true, for in that case the stake will certainly not be lost, and to offer any non-zero gain if E is true would be unfair to the bookmaker. There is no maximum possible fair odds, but we can think of infinite odds being appropriate when E is considered to be certainly false, for then the stake will certainly be lost and any odds, however high, would be unfavourable to You. In general, a higher value of fair odds corresponds to a lower degree of belief in the truth of E. Therefore, in order to give the probability scale the required characteristics, our transformation must reverse the odds scale and also condense it.

Transforming odds to probability

A transformation of odds to probability must transform odds zero to probability one, infinite odds to probability zero, and higher values of odds must be transformed into lower probabilities.

Subject to these requirements, we could use any transformation we choose. Probability would be *defined* by this transformation, and its scale (i.e. the meaning of every probability from zero to one) determined by it. For example, an exponential transformation which defined $P(E|H)$ to equal $\exp\{-O(E|H)\}$ would meet the requirements. Letting E be the proposition of tossing Heads with a coin, we agreed that $O(E|H)=1$. The exponential transformation would define a scale of measurement on which the probability of tossing Heads would become $P(E|H)=e^{-1}=0.368$. We do not use this transformation because the scale it defines is not very convenient.

Scales of measurement are nearly always chosen so as to have an additivity property, that the measurement of two objects combined (in a suitable way) equals the sum of their individual measurements. Additivity is so commonplace that we take it for granted in everyday measurements like length and weight. In the next section we prove that an appropriate additivity property holds when we define probability as follows.

Fair odds definition of probability
Your probability for a proposition E given information H is defined to be

$$P(E|H) = \frac{1}{1+O(E|H)}.$$

In the case of a coin toss, for example, if Your fair odds is one as suggested, then Your probability is 0.5.

As a measurement technique, this approach is no better than direct measurement. It relies on You being able to specify Your fair odds value $O(E|H)$, which is a very similar task to direct measurement of $P(E|H)$. It does not lend itself to more accurate measurements, but the advantage of the fair odds approach is that it enables us to derive the two basic probability laws.

1.5 The Addition Law

There are three important ways of operating on propositions to define new propositions. The simplest is *negation*. The negation of a proposition E is the proposition $\neg E$, read 'not E', which is simply the proposition that E is false. E is true when $\neg E$ is false, and vice versa. Now consider two propositions, E and F. Two other propositions that are often of interest are the proposition that either E or F is true (or both), and the proposition that both are true. We denote these two propositions by $E \vee F$ (read 'E or F') and $E \wedge F$ (read 'E and F'). For instance, E might be the proposition that it rains today and F the proposition that it rains tomorrow (in a certain place). Then $E \vee F$ is the proposition that it rains either today or tomorrow, or both, and $E \wedge F$ is the proposition that it rains on both days. The proposition $E \wedge F$ is also known as the *conjunction* of E and F, while $E \vee F$ is their *disjunction*.

Operations on propositions

The negation of E is the proposition $\neg E$, read 'not E', which is true if E is false. The disjunction of E and F is the proposition $E \vee F$, read 'E or F', which is true if either E or F (or both) is true. The conjunction of E and F is the proposition $E \wedge F$, read 'E and F', which is true if both E and F are true.

We also introduce now another useful notation. It is tedious to keep saying 'Let E be the proposition that . . .', so we will henceforth use the following convention for defining propositions.

Defining propositions

The notation $E \equiv$ '. . .' defines E to be the proposition represented by the expression in quotation marks.

For instance, $E \equiv$ 'a coin lands Heads' means that E is the proposition that a coin lands Heads. The expression in quotation marks may be further abbreviated if there is no risk of ambiguity, so that if we have been considering a coin toss then $E \equiv$ 'Heads' could be used.

We now proceed to prove the first of the probability laws. Let E and F be two propositions whose conjunction $E \wedge F$ is certainly false, i.e. they cannot both be true. E and F are said to be *mutually exclusive*. Since we are using odds, a particularly appropriate example is a horse race. Suppose that $E \equiv$ 'a certain horse wins' and $F \equiv$ 'a different horse wins'. They cannot both win, especially if we agree to regard a dead heat between them as meaning that neither has won, and so E and F are mutually exclusive. We are about to place bets on each of the two horses, but the argument is phrased in quite general terms: E and F may be any two propositions which are mutually exclusive.

Your fair odds for E based on Your knowledge H is $O(E|H)$, and any bet made at odds $O(E|H)$ You would regard as fair. Likewise bets on F at odds $O(F|H)$ are fair for You. So suppose that You make the following two bets.

1. B_E is a bet on E at odds $O(E|H)$ and with stake £$(1 + O(F|H))$.
2. B_F is a bet on F at odds $O(F|H)$ and with stake £$(1 + O(E|H))$.

These curious stakes are chosen to make the combination of B_E and B_F rather interesting. The net result of these two bets depends on the truth or falsity of the propositions E and F: there are three cases to consider.

(a) If neither E nor F is true, You lose both bets. Your net loss is Your total stake £S, where

$$S = 2 + O(E|H) + O(F|H).$$

(b) If E is true but F is false, You win B_E but lose B_F. You gain £$O(E|H)(1+O(F|H))$ on B_E but lose £$(1+O(E|H))$ on B_F, so Your net gain is £G, where

$$G = O(E|H)O(F|H) - 1.$$

(c) If E is false but F is true, You win £$O(F|H)(1+O(E|H))$ on B_F but lose £$(1+O(F|H))$ on B_E. Your net gain is again £G.

You cannot win both bets because E and F are mutually exclusive.

The effect of making both these bets can now be seen as a bet on the proposition $E \vee F$, for if either E or F is true You gain an amount £G, and if neither is true You lose an amount £S. Since both B_E and B_F are fair bets, it is clear that for You this represents a *fair* bet on $E \vee F$, having stake £S and at odds

$$\frac{G}{S} = \frac{O(E|H)O(F|H) - 1}{2 + O(E|H) + O(F|H)}.$$

This, therefore, is Your fair odds for $E \vee F$, i.e. $O(E \vee F|H) = G/S$. It is now a simple matter to convert everything from odds to probabilities.

$$\begin{aligned}
P(E \vee F|H) &= \frac{1}{1 + O(E \vee F|H)} = \frac{1}{1 + G/S} = \frac{S}{S + G} \\
&= \frac{2 + O(E|H) + O(F|H)}{1 + O(E|H) + O(F|H) + O(E|H)O(F|H)} \\
&= \frac{1}{1 + O(E|H)} + \frac{1}{1 + O(F|H)} \\
&= P(E|H) + P(F|H).
\end{aligned}$$

This is the vital additive property of probabilities, which we express formally as the Addition Law.

The Addition Law

If propositions E and F are mutually exclusive, then

$$P(E \vee F|H) = P(E|H) + P(F|H). \tag{1.1}$$

{............

Example

You are contemplating bidding for an item of furniture in an auction, provided that You will not need to bid too high in order to buy it. Let $E \equiv$ 'You will not need to bid more than £20', and $F \equiv$ 'You will need to bid more than £20 but not more than £30'. You measure $P(E|H) = 0.4$ and $P(F|H) = 0.3$. Now E and F are mutually exclusive and $E \vee F =$ 'not more than £30'. Your probability for this proposition is given by the Addition Law as $P(E \vee F|H) = 0.7$.

............}

This example raises a very important question. Suppose that instead of deriving $P(E \vee F|H)$ by the Addition Law You were to obtain a value by direct measurement. Would Your three directly measured probabilities $P(E|H)$, $P(F|H), P(E \vee F|H)$ obey the Addition Law? In theory they should but in practice they may very well not. We discuss this matter in detail in Chapter 2.

A proposition E and its negation $\neg E$ are mutually exclusive. Furthermore, the disjunction $E \vee \neg E$ is a proposition ('E or not E') which is certain to be true. Therefore Your probability for this proposition is one. Thus,

$$1 = P(E \vee \neg E|H) = P(E|H) + P(\neg E|H)$$

using the Addition Law. Therefore $P(\neg E|H) = 1 - P(E|H)$.

Probability of a negation

For any proposition E,

$$P(\neg E|H) = 1 - P(E|H) . \tag{1.2}$$

Exercises 1(b)

1. Let W be the proposition of a nuclear war in the next decade. More precisely, let $W \equiv$ 'either the USA or the USSR drops a nuclear weapon on the other's territory at some time during the next ten years'. Let $A \equiv$ 'at some time in the future the USA 'strikes first', i.e. drops a nuclear weapon on the USSR before the USSR drops one on the USA', and let $R \equiv$ 'USSR strikes first'.

A person states the following as fair odds (for him, based on his current information H):

$$O(W \wedge A|H) = 20, \quad O(W \wedge R|H) = 15 .$$

Derive $P(W|H)$ using the definition of probability in terms of fair odds and the Addition Law.

2. You are about to toss two dice. Let $E_{10}\equiv$'the total score is greater than 10'. Let $F_{11}\equiv$'score equals 11' and $F_{12}\equiv$'score equals 12'. Measure your probabilities $P(F_{11}|H)$ and $P(F_{12}|H)$ and thereby determine a value for $P(E_{10}|H)$ using the Addition Law.

3. Following Exercise 2, let $E_9\equiv$ 'score exceeds 9' and $F_{10}\equiv$'score equals 10'. By measuring $P(F_{10}|H)$ obtain $P(E_9|H)$. Continue this process until you have derived a value for $P(E_1|H)$, where $E_1\equiv$'score exceeds 1'. Do you have $P(E_1|H)=1$?

4. (For mathematicians only.) Prove that no other definition of probability as a transformation of odds would yield the Addition Law.

1.6 The Multiplication Law

In this section we derive a law for Your probability for the proposition $E \wedge F$. E and F may be any two propositions. As in the last section, we will make a pair of fair bets, and it may help to think of E and F as propositions that two horses will win, but this time they run in different races. An added complication arises now: when You come to bet on the second race You know the result of the first race. Therefore, Your information changes. We shall often have to consider changes in information, and in doing so it is convenient to think of H as a proposition, namely the proposition that You would see/know/experience everything that You did actually see/know/experience since birth. Then when You learn the truth of proposition E Your information changes from H to $H \wedge E$. Or, if You learn that E is false it becomes $H \wedge \neg E$. Treating H as a proposition therefore has some advantages in our notation.

The proposition H

H may be operated on as a proposition. In particular, $H \wedge E$ represents Your information H plus the knowledge that E is true.

Consider two arbitrary propositions, E and F. Let Your initial information be denoted by H. Given H, Your fair odds for the proposition E is $O(E|H)$ and You make this fair bet:

B_E is a bet on E at odds $O(E|H)$ and with stake £1.

If E is false, You lose B_E and You make no further bets. But if E is true You make a second bet, this time on F. At this point You know $H \wedge E$. Given information $H \wedge E$ Your fair odds for F is $O(F|H \wedge E)$, and You bet Your original £1 stake plus all Your winnings in a fair bet on F.

$B_{F|E}$ is a bet on F at odds $O(F|H \wedge E)$ and with stake $\pounds(1 + O(E|H))$.

We now consider the net effect of this scheme (which, in horse racing terms, is known as a 'double').

(a) If E is false then You lose £1 from B_E. You do not make the bet $B_{F|E}$ so, whether or not F is true, £1 is Your total loss.

(b) If E is true but F is false You gain $\pounds O(E|H)$ from B_E but then You lose $\pounds(1 + O(E|H))$ from $B_{F|E}$. Your net loss is again £1.

(c) If E and F are both true then You win both bets. You win $\pounds O(E|H)$ from B_E and $\pounds O(F|H \wedge E)(1 + O(E|H))$ from $B_{F|E}$. Your total profit is $\pounds z$, where

$$z = O(E|H) + O(F|H \wedge E) + O(E|H)O(F|H \wedge E).$$

The overall effect is of a bet on the conjunction $E \wedge F$ with stake £1 and odds z. The two individual bets were fair and so this is also fair. It follows that z represents Your fair odds for $E \wedge F$. Therefore

$$P(E \wedge F|H) = \frac{1}{1+z} = \frac{1}{(1 + O(E|H))(1 + O(F|H \wedge E))}$$

$$= P(E|H)P(F|H \wedge E).$$

You may wonder why we say that z is Your fair odds for $E \wedge F$ given information H rather than $H \wedge E$. The reason is that this betting scheme as a whole is seen to be fair before You learn the truth of E, i.e. given only Your initial information H. To make the bet $B_{F|E}$ given only the information H would not be fair – given H the fair odds would be $O(F|H)$ – but we do not make the bet in this unqualified way. We know that $B_{F|E}$ will be made only if E is found to be true, and therefore it is clear from the start that the fair odds should be $O(F|H \wedge E)$.

We have now proved our second law.

The Multiplication Law

For any propositions E, F, H, where H represents Your information at some time,

$$P(E \wedge F|H) = P(E|H)P(F|H \wedge E). \tag{1.3}$$

The Addition and Multiplication Laws are the foundation of all probability theory. Many *theorems* will be proved in this book using these Laws. (The structure is hierarchical, later theorems being proved using earlier ones, but ultimately they all rest on the two Laws.) The role of the fair bets definition is purely to establish these Laws, and we will not have to refer to bets again. The result we proved at the end of the last section, equation (1.2), is actually our first theorem.

1.7 Independence

We introduced the Multiplication Law by imagining two horse races, and said that knowing the winner of the first race is extra information which may change Your fair odds for propositions concerning the second race. This is true in general, but in practice such information would not cause You to modify Your fair odds noticeably. Bookmakers consider the information value to be so slight that for the purposes of combined bets like the 'double' $(B_E, B_{F|E})$ they do not adjust the odds for the second bet at all. They would not, of course, claim that the result of one race gives no information at all about the result of another, simply that it gives so little information that it is not worth the trouble of explicitly recognizing it. There are several issues here that are dealt with fully in the next chapter, but it is useful to look at a kind of idealized formulation now.

Suppose that, given information H, if You were to learn that E is true Your probability for F would not change, i.e.

$$P(F|H \wedge E) = P(F|H) . \tag{1.4}$$

Then we say that F *is independent of* E *given* H. Strictly speaking, we should not use the word 'independent' without qualification, because there are so many different kinds of independence that scientists find useful. The kind we are referring to is more correctly known as statistical independence, or stochastic independence. (The word 'stochastic' means 'relating to random phenomena'.) For instance, there is a weaker kind of independence known as logical independence, whereby to say that F is (logically) independent of E given H means that on the basis of information $E \wedge H$ the proposition F is neither logically true nor logically false, i.e. it is still possible for it to be either. However, in this book we shall use statistical independence so frequently that it would be irritating to give it its full title on every occasion. Therefore, the word 'independent' on its own always means 'statistically independent'.

From (1.3) and (1.4) together we find an equivalent representation for F being independent of E given H,

$$P(E \wedge F|H) = P(E|H) P(F|H) \tag{1.5}$$

from which, in turn we find

$$P(E|H) = P(E|H \wedge F)$$

by using (1.3) with the roles of E and F interchanged. We now see that if F is independent of E given H then E is independent of F given H. Independence is a symmetric relationship – the knowledge of either does not affect Your probability for the other. Because of this symmetry it is usual to define independence using the symmetric form (1.5).

Definition of independence

E and F are said to be (statistically) independent given H if

$$P(E \wedge F | H) = P(E | H) P(F | H) .$$

It is important to bear the condition 'given H' in mind, because independence is always relative to the information You have. E and F may not be independent when referred to a different body of information H_1.

Exercises 1(c)

1. A bag contains 4 balls, identical except for colour. Three balls are red and one is white. The bag is shaken and a ball taken out: it will then be shaken again and a second ball taken out. Let $R_1 \equiv$ 'first ball is red', $W_1 = \neg R_1 =$ 'first is white', $R_2 \equiv$ 'second ball is red', and $W_2 \equiv \neg R_2$. By direct measurement, of probabilities such as $P(R_1 | H)$ and $P(R_2 | R_1 \wedge H)$, and the Multiplication Law, obtain the first four probabilities below. Then derive the remaining probabilities using the Addition and Multiplication Laws.

- (i) $P(R_1 \wedge R_2 | H)$
- (ii) $P(R_1 \wedge W_2 | H)$
- (iii) $P(W_1 \wedge R_2 | H)$
- (iv) $P(W_1 \wedge W_2 | H)$
- (v) $P(R_2 | H)$
- (vi) $P(W_2 | H)$
- (vii) $P(R_1 | R_2 \wedge H)$
- (viii) $P(W_1 | R_2 \wedge H)$
- (ix) $P(R_1 | W_2 \wedge H)$

2. A bag contains two balls, one red and one white. The following scheme is known as 'Polya's urn'. Balls are drawn out successively, the contents of the bag changing after each draw according to the rule: 'Put the ball that was drawn out back into the bag and add another ball of the same colour'. Thus the number of balls in the bag increases after each draw. By repeated application of the Multiplication Law, find your probabilities $P(E_k | H)$ for $k = 1, 2, \ldots, 10$, where $E_k \equiv$ 'the first k balls drawn are all red'.

3. Let $L_1 \equiv$ 'snow will fall in central London on 1st January next year', and $L_2 \equiv$ 'snow will fall in central London on 2nd January'. Using the Multiplication Law and suitable measured probabilities, derive a value for your probability for the proposition $L_1 \wedge L_2$.

4. Consider each possible pairing of propositions from the following list, and decide which pairs of propositions you would regard as independent given your current information. NPUS is short for 'the Next President of the United States'.

$D \equiv$ 'NPUS will be a Democrat'.

$B \equiv$ 'NPUS will be black'.

$N \equiv$ 'NPUS will not be a US citizen'.

$S \equiv$ 'NPUS will be a smoker'.

Think of several other pairs of propositions which you would regard as independent.

5. Can mutually exclusive propositions be independent?

6. Repeat Exercise 1 with the following alteration: instead of simply removing the first ball from the bag, the rule of Polya's urn (Exercise 2) is applied after the first ball is drawn.

2

Probability measurements

In this chapter we resolve two important issues which we raised in chapter 1. The first is the fact that probabilities obtained by direct measurement may not in practice obey the probability laws. Section 2.1 discusses true probabilities and measured probabilities, and considers the significance of the failure of measured probabilities to follow the probability laws. A new form of measurement, elaborative measurement, is introduced in Section 2.2. Elaboration is the key to all the techniques presented later.

The second issue, the equivalence of the two definitions of probability, is resolved in Section 2.6. Sections 2.3 to 2.5 present a series of theorems which are needed for this resolution. Section 2.3 extends the Addition Law in two ways: the disjunction theorem provides an expression for $P(E \vee F | H)$ when E and F are not mutually exclusive, and is used to show that the Addition Law applies under the weaker condition that $P(E \wedge F | H) = 0$. The Law is extended again in Section 2.4 to the disjunction of more than two propositions. Section 2.5 concerns partitions. A partition is a set of propositions one and only one of which must be true. The partition theorem states that Your probabilities for the propositions in a partition sum to one. The case where the probabilities are all equal is of particular interest, and is shown in Section 2.6 to be the basis of very precise measurement in certain contexts.

2.1 True probabilities

Nothing can be measured with perfect accuracy. There is a difference between the true value and the value we get when using a measuring device. Most of the time the distinction is so unimportant that we are barely aware of it. We are accustomed to having *sufficiently* accurate measuring devices available for all our everyday measurements. The fact that when I measure my waist circumference with a tape measure I can expect an error of perhaps a centimetre is not a serious problem. If I could only measure it to within an accuracy of *ten* centimetres then I would have difficulty buying trousers. We are not used to a situation where the available measuring devices do not provide sufficient accuracy for the task at hand. Yet it is precisely this problem that we were faced with in Chapter 1 when trying to measure probabilities.

Consider the proposition $E \equiv$ 'white Christmas' of Section 1.2. If I wished to measure my probability $P(E | H)$ to an accuracy of 0.1 by direct measurement I should have to make decisions like 'E is more likely than $R(7,93)$ but less likely than $R(8,92)$'. I cannot make such fine comparisons with any confidence. Even the comparisons I think I can make may be illusory. I may, for instance decide today that E is more probable than $R(1,19)$ but tomorrow that $R(1,19)$ is more probable than E. Christmas is some months away as I write, so my information H will not change in any relevant sense between today and tomorrow.

The fault does not lie in direct measurement but in ourselves. Suppose that You were some kind of superhuman, capable of making arbitrarily fine comparisons and with perfect reasoning powers. Then You could measure Your probabilities perfectly accurately, even using direct measurement. There would be no limit to the precision of Your judgements, so You could make statements like 'Given my information H, I believe that proposition E is more probable than $R(993,247)$ but less probable than $R(994,247)$'. And You would mean precisely that: You are not going to change Your mind unless Your information changes. Giving You these superhuman abilities turns You into a perfect measurer of Your probabilities.

True probability
Your true probability $P(E|H)$ is the value You would obtain for it by direct measurement if You were capable of arbitrarily fine comparisons and possessed perfect reasoning powers.

We pause to emphasize that by referring to a probability as 'true' we mean only that it is Your true value. It is the value that You would give, based on Your information, if You had superhuman powers of thought. People will always have different information, and therefore their true probabilities will differ.

Your true probability for a proposition E given information H is unique. No amount of further thought could produce a different value because You have already applied Your hypothetical perfect reasoning powers, and have thereby taken full account of all Your information. But to Your one true probability $P(E|H)$ correspond many different *measured probabilities*. Each different measuring technique You use will produce a new measurement. You are also likely to get different results if You remeasure it using the same technique. According to our definition above, $P(E|H)$ represents Your true probability for E given H. When we require symbols for measured probabilities we will add a suitable subscript to P. Thus $P_1(E|H)$, $P_A(E|H)$, $P_+(E|H)$ will all be measured probabilities, the results of three different attempts to measure Your true probability $P(E|H)$.

Measured probability
True probabilities are unattainable in practice. Corresponding to the true probability $P(E|H)$ may be many measured probabilities, such as $P_1(E|H)$, identified by subscripts on the letter P.

The Addition and Multiplication Laws are statements about true probabilities, for in deriving the laws we used certain reasoning processes. *If* certain bets were fair individually, *then* in combination they implied that a third bet was fair, and hence a relationship was established between probabilities. With perfect reasoning powers You could not fail to notice these consequence. So Your true probabilities would naturally follow these laws and also any theorems, known or unknown, which could be proved from them. Measured probabilities, on the other hand, are necessarily the result of imperfect thought, and so it is quite possible that they might break the laws of probability.

Coherence
Probability values which obey the Laws and theorems of probability are said to be coherent. True probabilities are necessarily coherent, but measured probabilities are typically non-coherent.

Consider a very simplified example. An archaeologist wishes to know whether a collection of pottery fragments all date from the same century. Let $S \equiv$ 'same century', and $D \equiv \neg S =$ 'two or more different centuries'. Based on the archaeologist's information H, which contains substantial technical knowledge about the dating of pottery, she considers the proposition S and compares it with the reference propositions $R(r,w)$. Her direct measurement is

$$P_1(S|H) = 0.7 .$$

She now considers each fragment in turn, asking herself whether it might come from a different century from the rest, and in this way compares D with the propositions $R(r,w)$, obtaining

$$P_1(D|H) = 0.45 .$$

Since these two probabilities do not sum to one they are non-coherent. If this fact is pointed out to the archaeologist, what should she do?

Her measured probabilities are represented by the point M in Figure 2.1, overleaf, having coordinates 0.7 and 0.45. Her true probabilities must be represented by a point somewhere along the line AB which corresponds to the equation $P(D|H) = 1 - P(S|H)$, but of course her true probabilities are unknown, and so we do not know where on this line they lie. What is clear is that the point M cannot represent her true probabilities, and that it has incurred a certain amount of *measurement error* simply by being off the line AB. Suppose, for instance, that her true probabilities are represented by the point T in Figure 2.1. Then her measurement $P_1(S|H)$ has an error shown as x and $P_1(D|H)$ has an error shown as y. Wherever T is, her total measurement error $x+y$ is at least 0.15, which is the amount by which $P_1(S|H) + P_1(D|H)$ differs from one. This minimum total error occurs if T lies between the two points J and K in Figure

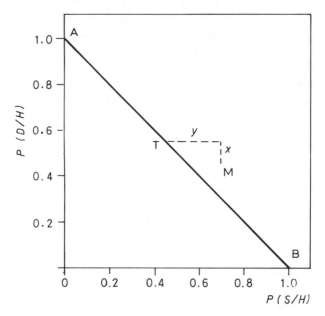

Figure 2.1. The archaeologist's non-coherence

2.2, which is a magnified view of the region around M.

There is obviously an opportunity for her to improve her measured probabilities, i.e. reduce their errors, by adjusting them so as to lie on the line. But where on the line? Not knowing her true probabilities, she might choose a point having even greater measurement errors than M. The answer is for her to think harder and more carefully about the two measurements. If true probabilities result from perfect reasoning then the way to achieve better measured probabilities is to use improved reasoning.

Overall she has assessed her probabilities too high, and will need to reduce one or both of them. If she decides to reduce both by the same amount she will obtain new measured probabilities

$$P_L(S\,|\,H) = 0.625\,,\quad P_L(D\,|\,H) = 0.375$$

represented by L in Figure 2.2. Or she might feel that the larger measurement is likely to have the larger error, and so decide to reduce them proportionately. This gives the point R with coordinates

$$P_R(S\,|\,H) = 0.609\,,\quad P_R(D\,|\,H) = 0.391\,,$$

lying on the line joining M to the origin. However, in practice our archaeologist is not likely to opt for either of these. Having directed her to consider her two measurements again she will probably feel that one of them is much more

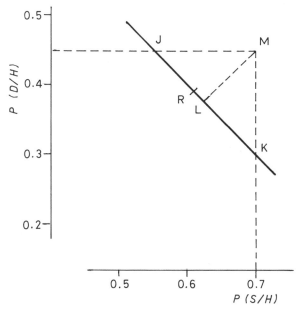

Figure 2.2. Removing the non-coherence

accurate than the other. Recalling the way in which she thought about the proposition D before, she may now decide that she gave too much weight to the very many ways in which the dates might differ. Her knowledge tells her that the pottery fragments really are very similar. So she leaves the measurement $P_1(S \mid H) = 0.7$ unchanged, and adjusts her measured probability for D to

$$P_K(D \mid H) = 0.3 ,$$

i.e. the point K in Figure 2.2. Like her original value $P_1(D \mid H)$ this probability is subject to measurement error. It may even be further from her true probability $P(D \mid H)$ than $P_1(D \mid H)$ was. However, because it is the result of more careful thought, it should be more accurate.

There are several useful lessons to learn from this example, some of which we shall develop more fully in the next section. The main points that concern the notion of coherence are quite simple. Whenever You discover that some of Your probability measurements fail to cohere, then by adjusting one or more of them so as to make them cohere You have a chance to improve them. Furthermore, the non-coherence focusses Your attention on particular ways in which they might be adjusted. As a general aid to good probability measurement, the search for and removal of non-coherence can be very valuable.

Exercises 2(a)

1. Using direct measurement, measure again your probability for the proposition that Shakespeare wrote more than 50 plays. Assuming the you have not obtained any extra relevant information since Exercise 1(a)3, this is a second measurement of the same true probability that you were attempting to measure then. Do you think that these measurements could be in error by as much as ± 0.2?

2. In Exercise 1(b)3, were your probability measurements coherent? If not, reconsider your probabilities $P(F_j|H)$.

3. A bag initially contains 5 red balls and 3 white balls. Successive draws are made from the bag according to the Polya urn rule (Exercise 1(c)2). Consider the first two draws and define R_1 and R_2 as in Exercise 1(c)1. First measure $P(R_1|R_2 \wedge H)$ directly and then derive a value using the method of Exercise 1(c)1. If the two measurements are different, resolve the non-coherence by altering one or more of your measurements.

4. A soccer match is about to be played between the two British teams Liverpool and Brighton. This match is in a knock-out competition and so if it results in a draw it must be replayed until one team has won. Let $L \equiv$ 'Liverpool win' and $B \equiv$ 'Brighton win'; in each case the victory may be after a replay. Let $F \equiv$ 'no replay is needed'. Criticize the following probability measurements.

$$P_m(L|H) = 0.7, \quad P_m(L \wedge F|H) = 0.45,$$
$$P_m(B \wedge F|H) = 0.35.$$

2.2 Elaboration

The idea of elaboration is based upon two simple precepts.

1. Some probabilities are easier to measure well than others.
2. It is possible to measure the more difficult probabilities by expressing them in terms of easier ones.

Our archaeologist found $P(D|H)$ more difficult to measure than $P(S|H)$. In thinking about D she found it hard to keep in perspective the many different alternatives, concerning which fragments dated from which centuries, whereas she found it rather easier to think about S. Therefore it was better, and more accurate, to measure $P(D|H)$ via her measurement $P_1(S|H)$ and the general result expressed in equation (1.2). Specifically, we can think of $P_K(D|H)$ arising as follows. There is a theorem,

$$P(D|H) = 1 - P(S|H), \tag{2.1}$$

which her true probabilities must follow. If she knew her true probability $P(S|H)$, she could use this result to *derive* her true probability $P(D|H)$. She does not know $P(S|H)$ but instead has a measurement of it, $P_1(S|H)$. If this is substituted for $P(S|H)$ in (2.1) the result will not be $P(D|H)$ but a measurement of it, so we define

$$P_K(D|H) \equiv 1 - P_1(S|H) .$$

$P_K(D|H)$ is subject to measurement error by virtue of the fact that $P_1(S|H)$ is. But since our archaeologist has rather more faith in the accuracy of $P_1(S|H)$ than $P_1(D|H)$, $P_K(D|H)$ should be more accurate, too.

We can obviously generalize this process.

Elaborative measurement

Let there be a theorem about Your true probabilities which asserts that a certain *target* probability $P(E|H)$ is equal to a given function

$$f \{P(E_1|H_1), \dots, P(E_n|H_n)\}$$

of other *component* probabilities $P(E_1|H_1), \dots, P(E_n|H_n)$. The target probability is said to have been *elaborated* in terms of the component probabilities.

Suppose that You have measurements $P_1(E_1|H_1), \dots, P_n(E_n|H_n)$ of the component probabilities. Then an *elaborative measurement* $P_e(E|H)$ of the target probability is

$$P_e(E|H) \equiv f \{P_1(E_1|H_1), \dots, P_n(E_n|H_n)\} .$$

Referring back to the two precepts with which we began this section, we can see that if $P(E|H)$ is difficult to measure well by direct measurement, but an elaboration can be found in terms of component probabilities which can be directly measured with reasonable accuracy, then $P_e(E|H)$ should be much better than direct measurement.

Measurement by elaboration is so important that we shall present two more examples here in order to make the process as clear as possible.

{...........

Example 1

Consider the toss of a coin. Let $T \equiv$ 'Tails'. This is a classic example of a proposition whose probability may be measured easily with great accuracy. Indeed, we have already remarked that anyone who is not irrational or superstitious will have fair odds for T very close to one, giving a probability close to $1/2$. Not exactly $1/2$, perhaps, because the two faces of a coin are different and it is plausible that the difference makes one face of the coin more likely to fall uppermost than the other. But it is hard to believe that the discrepancy is large, and certainly one's experience of coin-tossing seems to confirm that Heads and Tails are roughly equally likely, and that T is about as probable as $R(1,1)$. Therefore the measurements

$$P_1(T|H) = P_1(\neg T|H) = {}^1\!/_2$$

will be very close to the true probabilities for any reasonable person. (We assume that H represents the usual information available to people when tossing coins. It includes previous experience with coin-tossing, and does *not* include information that the coin or the manner of tossing it is abnormal in any way.)

Not all probabilities about coin-tossing are so easy to measure directly, but because they can be elaborated in terms of individual tosses they can still be measured accurately. Suppose we make ten successive tosses and let $K \equiv$ 'five Heads and five Tails'. It is by no means easy to measure $P(K|H)$ directly. Five Tails is the expected number but the vagaries of chance might very easily produce four or six Tails; perhaps even three or seven. Would you say that K is more or less probable than $R(1,3)$? Let $T_i \equiv$ 'Tails on the i-th toss'. The subscript i takes the values $1, 2, \ldots, 10$. The probabilities of the T_is can be given accurate measurements

$$P_1(T_1|H) = P_1(T_2|H) = \cdots = P_1(T_{10}|H) = {}^1\!/_2 . \tag{2.2}$$

Furthermore, knowing the results of previous tosses would not alter these measurements by any appreciable amount. This property is defined in Section 4.2 as 'mutual independence', and we show in Section 4.5 that $P(K|H)$ can be elaborated to obtain the measurement

$$P_e(K|H) = \frac{10 \times 9 \times 8 \times 7 \times 6}{5 \times 4 \times 3 \times 2 \times 1} \times \left(\frac{1}{2}\right)^{10} = \frac{252}{2^{10}}$$

$$= \frac{63}{256} = 0.246 . \tag{2.3}$$

By a small margin, this measured probability says that K is actually less probable than $R(1,3)$. But we must remember that this is only a measured probability and is subject to measurement error. Its error derives from errors in the component probabilities, which in this case are probabilities like (2.2).

So let us examine this error. Suppose that Your true probability for each T_i given H is p. The result (2.3) obtains if $p = \frac{1}{2}$, but in Section 4.5 it is shown that in general

$$P(K|H) = 252 p^5 (1-p)^5 . \tag{2.4}$$

Thus, if Your true value is $P(T_i|H) = 0.51$ then using (2.4) we find $P(K|H) = 0.246$ still, to three decimal places. If $p = 0.52$, then $P(K|H)$ falls to 0.244. We get the same figures if $p = 0.49$ and 0.48 respectively. Quite remarkably, the elaborative measurement (2.4) is actually more accurate than the component probabilities, such as (2.2), upon which it is based. In particular, since (2.4) never exceeds $252(1/2)^{10}$, You should definitely believe that K is less probable than $R(1,3)$.

...........}

{...........

Example 2

A doctor wishes to measure his probability for $A \equiv$ 'his patient suffers from the heart disease known as angina'. Having examined his patient he may still find it hard to assign a sensible probability value for A by direct measurement. To do so he must balance the fact that angina is not common against the fact that his examination has revealed certain symptoms that are suggestive of angina. This balance is nicely reflected in the elaboration

$$P(A \mid H_0 \wedge S)$$

$$= \frac{P(A \mid H_0) P(S \mid A \wedge H_0)}{P(A \mid H_0) P(S \mid A \wedge H_0) + P(\neg A \mid H_0) P(S \mid \neg A \wedge H_0)}, \quad (2.5)$$

where S denotes the symptoms that he has observed and H_0 denotes the rest of his information. (This is an example of Bayes' Theorem – see Section 3.2.) The doctor's present information H equals $H_0 \wedge S$. Now according to (2.5) the value of the target probability $P(A \mid H_0 \wedge S)$ depends on the relative magnitudes of $P(A \mid H_0) P(S \mid A \wedge H_0)$ and $P(\neg A \mid H_0) P(S \mid \neg A \wedge H_0)$. The fact that angina is uncommon means that if the doctor did not have the information contained in the observed symptoms S he would give a low probability to the proposition A. On the other hand, the fact that the symptoms point to angina means that those symptoms are much more likely to be seen in a patient with angina than in a patient without the disease.

Thus,

$$P(A \mid H_0) < P(\neg A \mid H_0) \quad (2.6)$$

but

$$P(S \mid A \wedge H_0) > P(S \mid \neg A \wedge H_0). \quad (2.7)$$

The relative magnitudes of the terms $P(A \mid H_0) P(S \mid A \wedge H_0)$ and $P(\neg A \mid H_0) P(S \mid \neg A \wedge H_0)$ depend on these two conflicting inequalities. If the doctor were to try to measure $P(A \mid S \wedge H_0)$ directly, he would have to weigh these magnitudes intuitively, just by thinking about them. The advantage of the elaboration (2.5) is that it expresses the conflict in terms of its two constituents (2.6) and (2.7). The four component probabilities $P(A \mid H_0)$, $P(\neg A \mid H_0)$, $P(S \mid A \wedge H_0)$, $P(S \mid \neg A \wedge H_0)$ are easier to think about. In fact, the doctor's training will have given him plenty of information (all contained in H_0) to enable him to measure these probabilities with reasonable accuracy directly. Given four measurements $P_1(A|H_0), \ldots, P_1(S|\neg A \wedge H_0)$ he then determines the corresponding elaborative measurement of his target probability

$P_2(A|S \wedge H_0)$

$$= \frac{P_1(A|H_0)P_1(S|A \wedge H_0)}{P_1(A|H_0)P_1(S|A \wedge H_0) + P_1(\neg A|H_0)P_1(S|\neg A \wedge H_0)} . \qquad (2.8)$$

..........}

Whenever the proposition whose probability You wish to measure is complicated, it is difficult to think clearly without breaking the problem down into smaller pieces. This is what elaboration does: it identifies the component probabilities and shows how they combine to produce the target probability. It is a basic scientific method, analogous to procedures common in other branches of science. A mathematician does not try to establish whether a theorem is true simply by thinking about it. Instead he prepares a step-by-step 'proof'. If each step of the proof is true and valid (which he must still decide by thinking about it – a kind of direct proof of each component) then the theorem is true. A physicist cannot weigh elementary particles by putting them on a balance. He has to use elaborative measurement, employing theory to express the target measurement in terms of other quantities which he can measure directly.

Exercises 2(b)

1. In Exercise 1(b)2, your value for $P(E_{10}|H)$ was obtained using the elaboration

$$P(E_{10}|H) = P(F_{11}|H) + P(F_{12}|H) .$$

Follow through your working in Exercise 1(b)3 to express your measurement of $P(E_1|H)$ as an elaboration in terms of all those probabilities which you measured directly.

2. Express your measurement of $P(R_1|R_2 \wedge H)$ in Exercise 1(c)1 as an elaboration in terms of directly measured probabilities. [Notice that you used exactly the same elaboration, but with different measurements of the component probabilities, in Exercises 1(c)6 and 2(a)3.]

3. Consider Exercise 1(a)4 and measure $P(C_2|H)$ by elaboration, using the Multiplication Law. Which do you consider more accurate in this case, direct or elaborative measurement?

4. Alan and Bernard play a game of shooting at a target. They take turns, Alan shooting first, and the first to hit the target is the winner. For simplicity, we will stop the game after Alan's second shot and if no shot has hit the target at that point we will declare Bernard the winner. Let $A \equiv$ 'Alan wins', $S_1 \equiv$ 'Alan's first shot hits', $S_2 \equiv$ 'Bernard's first shot hits', and $S_3 \equiv$ 'Alan's second shot hits'. Elaborate $P(A|H)$ in terms of $P(S_1|H), P(S_2|(\neg S_1) \wedge H)$ and $P(S_3|(\neg S_1) \wedge (\neg S_2) \wedge H)$.

Let $B \equiv$ 'Bernard wins', and elaborate $P(B|H)$ in terms of the same three component probabilities. Demonstrate that these two elaborations cohere (as everything concerning true probabilities must) in the sense that $P(A|H) + P(B|H) = 1$.

5. Let $E_1, E_2, \ldots, E_{10}$ be the propositions that on the next ten times that you go to work you will arrive late. Suppose that you regard these propositions as all having the same probability $p = P(E_i|H)$, and that they are independent (independence of more than one proposition is dealt with in Section 4.2). Then if $F \equiv E_1 \wedge E_2 \wedge \cdots \wedge E_{10} =$ 'you arrive late ten times in succession', it can be shown that $P(F|H) = p^{10}$. Measure your probability p, of being late on a single occasion, directly and then measure your probability $P(F|H)$ using the above elaboration. Why should this target probability be difficult to measure well directly?

[If you do not go to work regularly, or do not have a specific time for arrival, use some other appointment.]

2.3 The disjunction theorem

In Section 1.5 we introduced the concept of mutual exclusivity by saying that two propositions E and F were mutually exclusive if they could not both be true. If $E \wedge F$ is certainly false then $P(E \wedge F|H) = 0$. However, to say that a proposition 'cannot be true', that it is 'impossible' or 'certainly false', is stronger than to say that Your probability for that proposition given Your information H is zero. If a proposition A is genuinely impossible, i.e. it is *logically* false, then whatever Your information H Your true probability $P(A|H)$ will be zero – for no odds, however high, would make a bet on a logical impossibility fair. But it is not true that if $P(A|H) = 0$ then A is logically false. In fact, in Chapter 6 we shall encounter propositions which have probability zero and yet are still logically possible. Logical impossibility is objective, and in particular if A is impossible then $P(A|H) = 0$ for any H. It should therefore be distinguished from a subjective probability judgement that given Your information H Your (true) probability value for A is zero, i.e. $P(A|H) = 0$ for a particular H. Someone else, with different information H_d, may believe $P(A|H_d) \neq 0$.

In Section 1.5 we proved that the Addition Law, which states that $P(E \vee F|H) = P(E|H) + P(F|H)$, holds when E and F are logically mutually exclusive. In this section we demonstrate that the same result is true under the weaker condition that $P(E \wedge F | H) = 0$. We first obtain a result about $P(E \vee F | H)$ which holds quite generally.

The disjunction theorem

For any propositions E and F, and information H,

$$P(E \vee F | H) = P(E | H) + P(F | H) - P(E \wedge F | H).$$ (2.9)

{...........

Proof

Let $G_1 \equiv E \wedge F$, $G_2 \equiv E \wedge (\neg F)$, then $E = G_1 \vee G_2$ and $E \vee F = F \vee G_2$. Now $G_1 \wedge G_2$ is logically impossible, and so is $F \wedge G_2$. Therefore we can apply the Addition Law in each case to give

$$P(E | H) = P(G_1 | H) + P(G_2 | H),$$

$$P(E \vee F | H) = P(F | H) + P(G_2 | H).$$

Therefore

$$P(E \vee F | H) = P(F | H) + \{P(E | H) - P(G_1 | H)\}.$$

...........}

From the disjunction theorem it follows immediately that if $P(E \wedge F | H) = 0$ then $P(E \vee F | H) = P(E | H) + P(F | H)$. If E and F are *logically* mutually exclusive, by which we mean that $E \wedge F$ is logically impossible, then $P(E \wedge F | H) = 0$ for any H. It is in this circumstance that the Addition Law applies, but we have now generalized the result to the case where $P(E \wedge F | H) = 0$ for some specific H but not necessarily for all H. We then say that E and F are mutually exclusive *given H*.

Mutual exclusivity given H

E and F are said to be mutually exclusive given H if $P(E \wedge F | H) = 0$. If E and F are mutually exclusive given H then

$$P(E \vee F | H) = P(E | H) + P(F | H).$$ (2.10)

E and F are said to be logically mutually exclusive if $E \wedge F$ is impossible. If E and F are logically mutually exclusive then (2.10) holds for all H – this is the Addition Law.

Exercises 2(c)

1. A single card is to be drawn from a deck. Let $S \equiv$ 'Spade', $A \equiv$ 'Ace'. Measure your probabilities $P(S|H)$, $P(A|H)$, $P(S \vee A|H)$, $P(S \wedge A|H)$, and confirm that they cohere.

2. A hitch-hiker is waiting for a motorist to offer him a lift. Letting $F \equiv$ 'first car stops', $S \equiv$ 'second car stops', he measures $P_h(F|H) = P_h(S|H) = 0.2$, $P_h(F \wedge S|H) = 0.1$. Elaborate his probability $P(F \vee S|H)$.

3. A bag contains a large number of balls which are identical except for a number painted on each. For $i = 1, 2, \ldots, 6n$, there are i balls numbered i. (So, there is one ball labelled 1, two balls labelled 2, and so on.) One ball is to be taken out. Let $A \equiv$ 'its number is even' and $C \equiv$ 'its number is divisible by 3'. Given your information H, measure your probabilities $P(A|H)$, $P(C|H)$, $P(A \vee C|H)$, $P((\neg A) \vee (\neg C)|H)$ and $P(A \vee (\neg C)|H)$.

2.4 The sum theorem

We now need to extend our theory from results about two propositions to corresponding results about arbitrarily many propositions. When discussing many propositions we will use the subscript notation which will be familiar to many readers.

{............

Subscripts, repeated sums and products
When in algebraic work we need symbols for many different, but related, variables we use subscripts. The symbols $x_1, x_2, \ldots, x_n$ represent n quantities, and the symbol x_i represents any one of them, depending on which of the values $1, 2, \ldots, n$ the variable i has. The n variables are denoted collectively by $\{x_i\}$, or less formally by 'the x_is'.

The symbol Σ represents a repeated sum in the sense that $\sum_{i=1}^{n} x_i$ means $x_1 + x_2 + \cdots + x_n$. The range of values of i used in forming the sum are given by the expressions above and below the Σ, in this case '$i=1$' to 'n'. For typographical reasons they may be offset as in $\Sigma_{i=1}^{n}$, rather than $\sum_{i=1}^{n}$. The symbol Π means a repeated product in the same way – $\prod_{i=1}^{n} = x_1 \times x_2 \times \cdots \times x_n = x_1 x_2 \cdots x_n$. By altering the ranges of subscripting variables, using multiple subscripts, adding extra conditions and changing the expressions to be summed or multiplied, this notation can simplify a very wide variety of formulae. The following examples illustrate the main techniques.

$$\sum_{j=4}^{7} y_j x_{11-j} = y_4 x_7 + y_5 x_6 + y_6 x_5 + y_7 x_4.$$

$$\sum_{i=1}^{n} \prod_{j=1}^{m} a_{i,j} = (a_{1,1} a_{1,2} \cdots a_{1,m}) + (a_{2,1} a_{2,2} \cdots a_{2,m})$$

$$+ \ldots + (a_{n,1} a_{n,2} \cdots a_{n,m}) \, .$$

$$\sum_{j=1}^{3} \sum_{k=0}^{j} c_j^k = (1 + c_1) + (1 + c_2 + c_2^2) + (1 + c_3 + c_3^2 + c_3^3) \, .$$

$$\sum_{\substack{r=1 \\ r \neq s}}^{2n} \sum_{s=1}^{2n} z_{r+s} = (z_3 + z_4 + \ldots + z_{2n+1})$$

$$+ (z_3 + z_5 + z_6 + \ldots + z_{2n+2}) + \cdots$$

$$= 2z_3 + 2z_4 + 4z_5 + 4z_6 + \ldots + 2nz_{2n+1}$$

$$+ \ldots + 2z_{4n-1} \, .$$

..........}

The repeated disjunction $E_1 \vee E_2 \vee \cdots \vee E_n$ is represented by $\bigvee_{i=1}^{n} E_i$, and the repeated conjunction $E_1 \wedge E_2 \wedge \cdots \wedge E_n$ by $\bigwedge_{i=1}^{n} E_i$. These symbols are used in the same ways as the repeated sum and product symbols Σ and Π. For instance,

$$\bigwedge_{i=1}^{n} \bigvee_{j=0}^{1} F_{i,j} = (F_{1,0} \vee F_{1,1}) \wedge (F_{2,0} \vee F_{2,1}) \wedge \cdots \wedge (F_{n,0} \vee F_{n,1}) \, .$$

In this section we extend the Addition Law to the case of the disjunction of arbitrarily many mutually exclusive propositions. First we must extend our notion of mutual exclusivity to more than two propositions. The n propositions $E_1, E_2, E_3, \ldots, E_n$ will be said to be mutually exclusive if all pairs of them are mutually exclusive.

Many mutually exclusive propositions
A set of propositions $E_1, E_2, \ldots, E_n$ are said to be mutually exclusive given H if for every i and j between 1 and n, such that $i \neq j$, $P(E_i \wedge E_j | H) = 0$.

The case $n=2$ covers our earlier definition. The propositions will be logically mutually exclusive if they are mutually exclusive given any H.

The following simple theorem is useful generally, and in particular will be needed in proving the main result of this section.

The conjunction inequality
For any two propositions E and F, and information H,

$$P(E \wedge F | H) \leq P(E | H) \, . \tag{2.11}$$

{...........

Proof.

By the Multiplication Law $P(E \wedge F|H) = P(E|H)P(F|E \wedge H)$. Since probabilities are less than or equal to one, the right hand side of this expression is less than or equal to $P(E|H)$.

...........}

We may now state and prove the sum theorem, which generalizes the Addition Law to n mutually exclusive propositions.

The sum theorem

If the n propositions $E_1, E_2, \ldots, E_n$ are mutually exclusive given H then

$$P(\vee_{i=1}^{n}E_i|H) = \sum_{i=1}^{n} P(E_i|H).$$ (2.12)

{...........

Proof.

By induction. [*Mathematical Induction.* Suppose that a theorem may be stated in terms of an integer quantity n, and is known to be true in the case of $n = n_0$. To prove it true for all $n \geq n_0$, prove that if it is assumed to be true for $n_0 \leq n \leq k$ for any k then it is also true for $n = k+1$.] The theorem is trivially true for $n=1$, and in (2.10) we proved that it is true for $n=2$. Assume it true for $2 \leq n \leq k$, and consider $k+1$ propositions $E_1, E_2, \ldots, E_{k+1}$ which are mutually exclusive given H. For $i=1, 2, \ldots, k$, let $F_i \equiv E_i \wedge E_{k+1}$. Let $G \equiv \vee_{i=1}^{k} E_i$, so that $\vee_{i=1}^{k} F_i = G \wedge E_{k+1}$.

The first step is to prove that the F_is are mutually exclusive given H. For any i and j in the range $1, 2, \ldots, k$,

$$P(F_i \wedge F_j|H) = P(E_i \wedge E_j \wedge E_{k+1}|H) = 0$$

using (2.11) with $E \equiv E_i \wedge E_j$. The second step is to prove that G and E_{k+1} are mutually exclusive given H.

$$P(G \wedge E_{k+1}|H) = P(\vee_{i=1}^{k}F_i|H) = \sum_{i=1}^{k} P(F_i|H)$$ (2.13)

using the theorem in the case $n=k$, which we assumed to be true. Every term in the sum (2.13) is zero because the E_is are mutually exclusive given H. The third step is simply that

$$P(\vee_{i=1}^{k+1}E_i|H) = P(G \vee E_{k+1}|H)$$

$$= P(G|H) + P(E_{k+1}|H)$$ (2.14)

using (2.10). Finally,

$$P(G|H) = P(\vee_{i=1}^{k} E_i |H) = \sum_{i=1}^{k} P(E_i|H), \qquad (2.15)$$

using again the theorem in the case $n=k$, which is assumed true. The theorem for $n=k+1$ follows from combining (2.14) and (2.15).

...........}

We have repeatedly used the Addition Law in examples and exercises already. In some of these we could have simplified a little by using the sum theorem.

{...........

Example

A water engineer requires to measure her probability for $F\equiv$'failure of supply' to a village. She considers the various possible causes of failure, such as burst pipe, faulty valve or drought, and letting $C_i\equiv$'failure due to cause i', she measures

$$P_1(C_1|H) = 0.05, \qquad P_1(C_2|H) = 0.03,$$

$$P_1(C_3|H) = 0.01, \qquad P_1(C_4|H) = 0.03.$$

She considers that there are no other plausible causes of F, so that $F=\vee_{i=1}^{4}C_i$. Furthermore, although it is possible for more than one cause to occur at once, she regards the probability of this happening as negligible – $P_1(C_i \wedge C_j|H)=0$ for $i\neq j$. According to her measured probabilities, therefore, the C_is are mutually exclusive given H and she can use the sum theorem to elaborate her probability for F.

$$P_1(F|H) = \sum_{i=1}^{4} P_1(C_i|H) = 0.12.$$

...........}

Exercises 2(d)

1. A die is to be tossed, and your information H is such that you believe the die to be loaded in favour of higher scores. Letting $S_i\equiv$'score i', $i=1, 2,\ldots,6$, you measure the following probabilities.

$$P_1(S_1|H) = 0.10, \qquad P_1(S_2|H) = 0.14, \qquad P_1(S_3|H) = 0.16,$$
$$P_1(S_4|H) = 0.18, \qquad P_1(S_5|H) = 0.20, \qquad P_1(S_6|H) = 0.22.$$

By elaboration using the sum theorem, measure your probabilities for the propositions $L\equiv$'score 3 or less', $U\equiv$'score 3 or more', $E\equiv$'even', $O\equiv$'odd'.

2. The manager of soccer team X measures his probabilities for various possible scores in the forthcoming game against team Y. These are given in the table below. Assume that probabilities for either team scoring more than 3 goals are zero.

Goals for team Y	Goals for team X			
	0	1	2	3
0	.12	.11	.08	.04
1	.15	.10	.06	.02
2	.10	.08	.02	.01
3	.05	.04	.01	.01

Measure his probabilities for the following propositions:

$E_1 \equiv$ 'team X will score 1 goal';
$E_2 \equiv$ 'both teams will score at least 2 goals';
$E_3 \equiv$ 'the game will be drawn';
$E_4 \equiv$ 'team X will win';
$E_5 \equiv$ 'team Y will win by more than 1 goal'.

3. You are standing in the main shopping area of your town. You are going to stop the next person to walk past you and will ask him/her the following four questions.

Q1. Do you read a daily newspaper?
Q2. Do you know someone who was born in a Spanish-speaking country ?
Q3. Do you know someone who was born in Spain ?
Q4. Do you know someone who was born in Mexico ?

Assume that the person agrees to answer and, using only direct measurement, measure your probabilities for the following propositions concerning his/her answers. (Measure each probability separately, without reference to any other measurements.)

$A \equiv$ 'he/she reads a daily newspaper'.
$B \equiv$ 'he/she knows someone who was born in Spain but does not know anyone who was born in Mexico'.
$C \equiv$ 'he/she does not know anyone who was born in a Spanish-speaking country, and does not read a daily newspaper'.
$D \equiv$ 'he/she does not know anyone who was born in Mexico'.
$E \equiv$ 'he/she knows someone who was born in a Spanish-speaking country but does not know anyone who was born in Spain'.
$F \equiv$ 'he/she knows someone who was born in a Spanish-speaking country, but does not know anyone who was born in Mexico and does not read a daily

newspaper'.

$G \equiv$ 'he/she knows someone who was born in Spain and knows someone who was born in Mexico'.

$H \equiv$ 'he/she knows someone who was born in Mexico but does not know anyone who was born in Spain'.

$I \equiv$ 'he/she does not know anyone who was born in a Spanish-speaking country but reads a daily newspaper'.

$J \equiv$ 'he/she knows someone who was born in Spain but does not read a daily newspaper'.

$K \equiv$ 'he/she knows someone who was born in Mexico'.

$L \equiv$ 'he/she knows someone who was born in Mexico and knows someone who was born in Spain and reads a daily newspaper'.

$M \equiv$ 'he/she knows someone who was born in a Spanish-speaking country but does not know anyone who was born in Spain, and does not read a daily newspaper'.

Using only the sum theorem, examine your probability measurements for examples of non-coherence.

4. Prove that, for any E, F and information H, $P(E \vee F | H) \geq P(E | H)$.

5. 170 students were asked whether they had read the Bible (B), 'The Lord of the Rings' (L) or 'War and Peace' (W). 20 said they had read all three; 70 had read two but not all three; 35 had read none; 30 had read B and L but not W; 15 had read B and W but not L; 60 had read W; 85 had read B. What is your probability that a 'randomly chosen' student from this group has read 'The Lord of the Rings'?

2.5 Partitions

The sum theorem is useful because of its simplicity. An even simpler result is the partition theorem. First some definitions.

Exhaustivity

A set of propositions $E_1, E_2, \ldots, E_n$ are said to be exhaustive given H if

$$P(\vee_{i=1}^{n} E_i | H) = 1 .$$ (2.16)

Loosely speaking, the propositions are exhaustive if at least one of them must be true. More precisely, the probability of the proposition $F \equiv \neg \vee_{i=1}^{n} E_i$ is zero given H. If F is logically impossible then the E_is are exhaustive given any H, which we call logical exhaustivity. A partition is a set of propositions of which, again loosely speaking, one and only one must be true.

Partitions

A set of propositions are said to be a partition given H if (i) they are mutually exclusive given H, and (ii) they are exhaustive given H.

The partition theorem is now immediate from (2.12) and (2.16).

The partition theorem

If the propositions $E_1, E_2, \ldots, E_n$ form a partition given H then

$$\sum_{i=1}^{n} P(E_i | H) = 1 . \tag{2.17}$$

A logical partition is a set of alternatives, precisely one of which must be true, and it is extremely easy to think of examples. For instance if there are n candidates in a British parliamentary election then one and only one will be elected. Letting $E_i \equiv$ 'the i-th candidate on the ballot paper is elected' $(i=1, 2, \ldots, n)$, the E_is are clearly a logical partition. Or if a standard die is to be tossed and for $i=1, 2, 3, 4, 5, 6$ we let $E_i \equiv$ 'score i', then the E_is are again a logical partition. The simplest logical partition is with $n=2$, comprising a proposition E and its negation $\neg E$. The application of the partition theorem gives a result we proved in Section 1.5, that for any H, $P(\neg E | H) = 1 - P(E | H)$.

A set of mutually exclusive but not exhaustive propositions $E_1, E_2, \ldots, E_n$ can always be turned into a partition of $n+1$ propositions by adding one more proposition

$$E_{n+1} = \neg \bigvee_{i=1}^{n} E_i .$$

Another way of forming a logical partition is to start with any m propositions and construct all the $n = 2^m$ possible repeated conjunctions of those propositions and their negations. For example, with $m=2$ we start with two propositions E and F. Then the following $2^2 = 4$ propositions are a logical partition,

$$E \wedge F , \quad E \wedge (\neg F) , \quad (\neg E) \wedge F , \quad (\neg E) \wedge (\neg F) .$$

{..........

Example

In Section 1.2 we suggested that you measure probabilities for football matches. The propositions $W \equiv$ 'home win', $L \equiv$ 'away win' and $D \equiv$ 'draw' form a partition. (We ignore the possibility of cancellation.) Therefore for any H and any match,

$$P(W|H) + P(L|H) + P(D|H) = 1 .$$

If your measurements of these probabilities do not sum to one they are non-coherent. If you have met some probability theory before , then you will perhaps be aware of this result and will have automatically measured your probabilities coherently. If not, your probabilities may have been non-coherent: you should be able to improve your measurements by making them cohere. The same obviously applies to the propositions $R \equiv$ 'rise, $F \equiv$ 'fall', $S \equiv$ 'stay' for records in the pop charts.
..........}

To introduce a second theorem, we return to the example of tossing a die. Unless You have exceptional information, You will have no reason to believe that any particular score is any more probable than any other. This situation, where symmetries in the propositions forming a partition lead You to assign them equal probabilities, arises so frequently that we present the following theorem.

The equi-probable partition theorem
If $E_1, E_2, \ldots, E_n$ are a partition given H, and if given H they have equal probabilities, then

$$P(E_1|H) = {}^1/n , \quad P(E_2|H) = {}^1/n ,$$

$$\ldots, \quad P(E_n|H) = {}^1/n . \tag{2.18}$$

{..........
   Proof.
   This is a trivial consequence of the partition theorem.
..........}

The traditional equipment of gambling games provide many instances of equi-probable partitions. We have already mentioned coins and dice; other examples are cards and roulette wheels. If a standard deck of cards is well shuffled and a single card drawn from it then there is a logical partition of 52 propositions, that the card drawn will be each of the 52 cards in the deck, and unless Your information is very unusual You will judge them equi-probable. Thus Your probability for any specific card being drawn is $^1/52$.

2.6 Symmetry probability

Probability theory began in the eighteenth century with the study of gambling games. Equi-probability seemed so obvious to its pioneers that the earliest definitions of probability assumed it. If an experiment could result in one of n

(equi-probable) different outcomes, and a proposition E asserted that the outcome would be one of a specified r outcomes, then the probability of E was defined to be r/n. It was quickly perceived that this approach failed as soon as one needed to measure probabilities in contexts where obvious symmetries could not be seen. Nowadays we adopt a more methodical approach, and the definitions of those early mathematicians become a simple theorem.

The symmetry probability theorem

If a proposition E states that one of a specified r propositions is true, and if those r propositions are members of a set of n propositions which comprise an equi-probable partition given H, then

$$P(E|H) = r/n .$$

{..........

Proof.

Let the r propositions be denoted by $F_1, F_2, \ldots, F_r$, so that $E = \bigvee_{i=1}^{r} F_i$, and let the n propositions of the partition be $F_1, F_2, \ldots, F_n$. By the equi-probable partition theorem $P(F_i|H) = 1/n$ for each i. Since the F_is are a partition they are mutually exclusive given H, therefore the r propositions which make up E are also mutually exclusive given H. Using the sum theorem,

$$P(E|H) = \sum_{i=1}^{r} P(F_i|H) = r/n .$$

..........}

For example, if a single card is drawn from a deck of cards, the probability that it is a heart is $13/52 = 1/4$, and the probability that it is an ace is $4/52 = 1/13$.

In discussing problems involving dice, cards and coins we have been asserting probabilities without proper regard for the distinction between true and measured probabilities. We must now reexamine those assertions. Consider again the toss of a die, where $E_i \equiv$ 'score i'. The judgement that the E_is are equi-probable seems natural, but can we assert that Your true probabilities are equal? First we must recognize that Your information H may make equi-probability inappropriate. You may have special information about the die, or the way in which it will be tossed, that suggests that certain scores are more likely than others. In particular, You may have observed many tosses of this die already in which sixes occurred much less frequently than ones. However, unless You have this kind of exceptional information, equi-probability does seem appropriate.

Even excluding special information, equi-probability is still a judgement, and the resulting probabilities are measurements. The point is that those measurements are highly accurate. Your true probability $P(E_6|H)$ may not be precisely $1/6$ but, without strong information to the contrary, it will surely be very close to $1/6$.

In examples based on coins, dice and cards, the symmetry of the equipment argues for an equi-probability judgement. In all these cases, provided we exclude from H any unusual or special information, then anyone who refused to judge the various possible outcomes as equi-probable would be thought irrational, superstitious or simply eccentric by the vast majority of other people. In such cases, equi-probability has an air of objective validity rather than a subjective judgement. Strictly speaking, subjectivity has not disappeared. People's true probabilities for the proposition E_6, of throwing a six with a single die toss, may not all be precisely equal, since each person may have a true probability not precisely one-sixth, but the variation between people is very small. This can be ignored in practice because the differences will not be amenable to measurement. Therefore it is sensible to assert that $P(E_6|H)=1/6$, i.e. that Your true probability of E_6, based on Your information H, is one-sixth, and to make this assertion for all people (except possibly those whose information H is special). We call this an *objective probability*.

Objective probabilities
In contexts such as gambling games, equi-probability is suggested by symmetry. Probability measurements obtained by the symmetry probability theorem are highly accurate. In such cases we regard those measurements as essentially true, and refer to them as objective probabilities.

Another example of equi-probability is drawing balls from a bag, as in Section 1.2. We described a bag containing $r+w$ balls, from which one ball was to be drawn. Equi-probability is natural, and indeed we tried to formulate the draw so that this judgement was inevitable. Your probability for any specific ball being drawn is therefore $1/(r+w)$. Since r of the balls are red, the symmetry probability theorem can be applied to show that $P(R(r,w)|H)=r/(r+w)$. We have therefore delivered the promised justification of the reference probabilities which were asserted in Section 1.2. And since our whole development is based on the definition of probability through fair odds in Section 1.4, this proof reconciles the two definitions of probability – by fair odds and by direct measurement.

In Section 2.2 we studied elaboration and said that a good elaboration uses component probabilities which can be measured accurately. The special context of gambling games presents us with the ultimate in measurement accuracy for component probabilities – objective probabilities. As a result, very complex target probabilities can also be measured objectively. Now in practical contexts we rarely have the luxury of being able to use objective probabilities like these. Nevertheless, good elaboration serves to identify components which can be measured with relatively high degrees of objectivity, and to separate these from more subjective components where different people's information may lead them to assign very different measurements. A very important example of this kind of elaboration is studied in Chapter 8.

Objective probabilities such as arise in gambling games have a number of worthwhile applications, as well as providing a convenient source of examples to illustrate theoretical results. They will therefore be used extensively throughout this book. It would be tedious and pedantic to remind the reader on all such occasions that Your information H must not be special, and that probabilities obtained by application of the symmetry probability theorem are not true probabilities but highly accurate measurements for which that theorem acts as an elaboration. We shall instead assume equi-probability wherever the context of the example suggests it, and treat resulting probabilities as true. The reader should tackle exercises in the same spirit: assume equi-probability where it is obviously appropriate, and treat resulting probabilities as true probabilities.

{...........

Example

Two dice, one white and one yellow, are to be tossed. Let $E_{i,j} \equiv$ 'score i on white, j on yellow' for $i,j=1, 2, \ldots, 6$. The 36 $E_{i,j}$s form an equi-probable partition. Now consider another partition, $S_k \equiv$ 'total score k' for $k=2, 3, \ldots, 12$. The probabilities for the S_ks are readily obtained from the symmetry probability theorem. For instance, a total score of 2 can only arise from $E_{1,1}$, therefore $P(S_2|H) = 1/36$. A total score of 5 can arise from $E_{1,4}$ or $E_{2,3}$ or $E_{3,2}$ or $E_{4,1}$, therefore $P(S_5|H) = 4/36 = 1/9$. The full set of probabilities,

$$P(S_2|H) = P(S_{12}|H) = 1/36,$$

$$P(S_3|H) = P(S_{11}|H) = 2/36 = 1/18,$$

$$P(S_4|H) = P(S_{10}|H) = 3/36 = 1/12,$$

$$P(S_5|H) = P(S_9|H) = 4/36 = 1/9,$$

$$P(S_6|H) = P(S_8|H) = 5/36,$$

$$P(S_7|H) = 6/36 = 1/6,$$

are easily obtained, and can also be seen to satisfy the partition theorem.

...........}

Exercises 2(e)

1. Confirm that the probabilities given in Exercises 2(d)1 and 2(d)2 satisfy the partition theorem.

2. Define the $2^4 = 16$ propositions in the logical partition obtained from the person's four answers to the four questions in Exercise 2(d)3. (For instance, the first proposition might be that the person answers No to all four questions.) Reduce these to a logical partition of 10 propositions by removing 6 combinations of answers which are logically impossible. Express each of the propositions A to M as a disjunction of one or more of the 10 propositions in this partition. Does this exercise help you to discover more non-coherences in your measured probabilities?

3. Consider the birthday of the next stranger you will meet. Assume an equiprobable partition for the propositions that his/her birthday is on each of the 365 days of the year (ignoring leap-years). Define the partition of 12 propositions corresponding to the month of his/her birthday and measure your probabilities for this partition. What probability would you give to the proposition that he/she was born later in the year than you?

4. Consider two tosses of a coin. Define a partition of 3 propositions:

$E_1 \equiv$ 'first is Heads';
$E_2 \equiv$ 'first is not Heads, but second is';
$E_3 \equiv$ 'neither is Heads'.

An early probabilist thought that these propositions were equi-probable and so gave each the probability $1/3$. Alternatively, define:

$F_1 \equiv$ 'both are Heads';
$F_2 \equiv$ 'only one is Heads';
$F_3 \equiv$ 'neither is Heads'.

Some people may regard this as an equi-probable partition. Under what condition could these two sets of probability judgements be coherent with each other? Criticize both sets of probabilities by reference to your own equi-probable partition.

3
Bayes' theorem

This chapter is concerned with two very important elaborations. In Section 3.1 we consider 'extending the argument', an elaboration which can be used whenever You feel that a target probability would be easier to measure if You had some extra information. Bayes' theorem, which we derive in Section 3.2, can be considered as having the opposite effect. It is useful when You feel You could measure probabilities accurately if some specific information were excluded from H. Its primary function is to describe how Your probabilities are affected by gaining new information, and this role is explored in Section 3.3. Section 3.4 considers the effect on Bayes' theorem when one or more of its component probabilities are zero. It shows the dangers of the kind of prejudice that unreasonably asserts a proposition to be impossible. Finally, in Section 3.5 we examine a real application in some detail.

3.1 Extending the argument

A student sits an examination with multiple-choice questions. Consider a single question, where he must indicate which of four stated answers he believes to be correct. Let $C \equiv$ 'he chooses the right answer'. You wish to measure $P(C|H)$. You decide that he is a poor student, and Your measured value for the probability that he knows the right answer is only 0.3. But there is a difference between C and $K \equiv$ 'he knows the right answer'. Under examination conditions he might mistakenly give a wrong answer even if he knows the right one, but this is unlikely. Suppose that Your two measurements so far are

$$P_1(K|H) = 0.3, \quad P_1(C|K \wedge H) = 0.95 .$$

Therefore, using the Multiplication Law as an elaboration, $P_1(C \wedge K|H) = 0.3 \times 0.95 = 0.285$. You still do not have a value for $P(C|H)$, but You can complete the elaboration by

$$P(C|H) = P(C \wedge K|H) + P(C \wedge (\neg K)|H) .$$

You must now consider his chances of circling the right answer by luck. Given $(\neg K) \wedge H$, suppose he guesses, with a probability 0.25 of choosing correctly. Then $P_1(C \wedge (\neg K)|H) = 0.7 \times 0.25 = 0.175$. So $P_1(C|H) = 0.285 + 0.175 = 0.46$. The complete elaboration is

$$P(C|H) = P(K|H)P(C|K \wedge H) + P(\neg K|H)P(C|(\neg K) \wedge H) ,$$

which is a simple case of the following general theorem.

Extending the argument

Let $E_1, E_2, \ldots, E_n$ be a partition given H. Then for any proposition F,

$$P(F|H) = \sum_{i=1}^{n} P(E_i|H) P(F|E_i \wedge H) . \qquad (3.1)$$

{...........

Proof.

First consider the propositions $G_i \equiv F \wedge E_i$, $i = 1, 2, \ldots, n$, and let $B \equiv \bigvee_{i=1}^{n} E_i$. Note that

$$\bigvee_{i=1}^{n} G_i = \bigvee_{i=1}^{n} (F \wedge E_i) = F \wedge (\bigvee_{i=1}^{n} E_i) = F \wedge B .$$

Now since the E_is are exhaustive given H, $P(\neg B|H) = 0$. Therefore, using the conjunction inequality (2.11), $P(F \wedge (\neg B)|H) = 0$. Since $P(F|H) = P(F \wedge B|H) + P(F \wedge (\neg B)|H)$, we have

$$P(F|H) = P(F \wedge B|H) = P(\bigvee_{i=1}^{n} G_i|H) .$$

We next notice that the G_is are mutually exclusive given H because $G_i \wedge G_j = (F \wedge E_i) \wedge (F \wedge E_j) = F \wedge (E_i \wedge E_j)$ and because the E_is are mutually exclusive given H (using (2.11) again). Therefore from the sum theorem,

$$P(F|H) = P(\bigvee_{i=1}^{n} G_i|H) = \sum_{i=1}^{n} P(G_i|H) .$$

Applying the Multiplication Law to each of the probabilities $P(G_i|H) = P(F \wedge E_i|H)$ completes the proof.

...........}

Extending the argument is one of the most useful elaborations. It arises naturally whenever, in answer to the question 'What probability would you give to the proposition F?', one is tempted to begin one's reply with 'It depends on $\cdots$'. If the knowledge that some proposition E_i in a partition $\{E_1, E_2, \ldots, E_n\}$ is true would make it easier to measure Your probability for F, then this shows that the component probabilities $P(F|E_i \wedge H)$ in (3.1) are measurable more accurately than the target probability $P(F|H)$. If the other components $P(E_i|H)$ can also be measured well, the elaboration will be successful. We continue this section with three more examples of extending the argument.

{...........

Example 1

A bag contains one white ball. A die is tossed and if its score is i then i red balls are put into the bag. The bag is shaken and one ball taken out. Letting

$R \equiv$'ball is red' what is $P(R|H)$? Direct measurement is inadvisable here because by extending the argument to include the toss of the die we can achieve objective probabilities. Let $S_i \equiv$'score i on die'. Then for $i = 1, 2, \ldots, 6$,

$$P(S_i|H) = \frac{1}{6}, \quad P(R|S_i \wedge H) = \frac{i}{i+1},$$

$$\therefore P(R|H) = \sum_{i=1}^{6} P(S_i|H) P(R|S_i \wedge H)$$

$$= (\tfrac{1}{6}) \times (\tfrac{1}{2}) + (\tfrac{1}{6}) \times (\tfrac{2}{3}) + \ldots + (\tfrac{1}{6}) \times (\tfrac{6}{7})$$

$$= \frac{1851}{2520} = 0.734 .$$

..........}

{..........

Example 2

My hi-fi is making irritating, buzzing sounds on loud passages. If it costs more than £25 to fix then I cannot afford it this month. Let $T \equiv$'it costs more than £25'. What is $P(T|H)$? By direct measurement I arrive at the value $P_d(T|H) = 0.6$, but I would find it much easier to think about the probable cost if I knew what kind of fault my equipment was suffering from. So I define a list of possible faults and associated propositions:

$F_1 \equiv$ 'damage to speakers',
$F_2 \equiv$ 'foreign matter in amplifier',
$F_3 \equiv$ 'damage to amplifier circuit',
$F_4 \equiv$ 'dust on stylus',
$F_5 \equiv$ 'damage to pickup',
$F_6 \equiv$ 'faulty connection between units',
$F_7 \equiv$ 'something else'.

This is not a logical partition. Proposition F_7 makes them logically exhaustive (and such a 'catch-all' proposition is frequently used for this purpose), but they are not logically mutually exclusive. However, I regard the chance of two or more distinct faults being present as very small. So I begin by asserting that $P_1(F_i \wedge F_j|H) = 0$ for all i,j pairs; the F_is are (measured as) a partition given H. I now extend the argument to include possible faults, by measuring the following probabilities.

$P_1(F_1|H) = 0.1$,	$P_1(T|F_1 \wedge H) = 1.0$,
$P_1(F_2|H) = 0.2$,	$P_1(T|F_2 \wedge H) = 0.2$,
$P_1(F_3|H) = 0.25$,	$P_1(T|F_3 \wedge H) = 0.7$,
$P_1(F_4|H) = 0.05$,	$P_1(T|F_4 \wedge H) = 0.0$,
$P_1(F_5|H) = 0.1$,	$P_1(T|F_5 \wedge H) = 0.2$,

$$P_1(F_6|H) = 0.1 \, , \qquad\qquad P_1(T|F_6 \wedge H) = 0.2 \, ,$$
$$P_1(F_7|H) = 0.2 \, , \qquad\qquad P_1(T|F_7 \wedge H) = 0.6 \, .$$

Finally, I apply the theorem as an elaboration to arrive at a new measurement of the target probability.

$$P_1(T|H) = (0.1 \times 1.0) + (0.2 \times 0.2) + (0.25 \times 0.7) + \ldots + (0.2 \times 0.6)$$

$$= 0.439 \, .$$

This is, of course, not coherent with my direct measurement $P_d(T|H) = 0.6$. On reflection, I think that my direct measurement was too pessimistic. The elaboration shows me that there are a number of quite plausible explanations of the buzzing noise that should be cheap to repair. I feel that $P_1(T|H) = 0.439$ is the more accurate figure, but there is still measurement error, which suggests rounding to $P_2(T|H) = 0.45$.

............}

{...........

Example 3

Our third example is a gambling game, and we assume objective probabilities. It is a complex example requiring a series of elaborations, and shows that we already have enough probability theory to handle quite substantial problems.

The game of 'craps' is played between a Player and the Banker. The Banker rolls two dice. He wins immediately if he scores '7, 11 or doubles', i.e. a total of 7 or 11, or both dice the same. If not, his score is noted (say x) and the dice pass to the Player. He rolls them and wins if he scores '7, 11 or doubles'. The Banker wins if the Player's total equals x. Otherwise the Player rolls the dice again. He keeps rolling until either he rolls a total of x (Banker wins) or he gets '7, 11 or doubles' (Player wins). Our problem is to obtain the probability of $W \equiv$ 'Player wins'. We first elaborate by extending the argument to include the result of the Banker's roll. Define a logical partition $E_2, E_3, \ldots, E_{11}$.

$E_2 \equiv$ 'doubles',	$E_3 \equiv$ 'score 3',
$E_4 \equiv$ 'score 4 but not double 2',	$E_5 \equiv$ 'score 5',
$E_6 \equiv$ 'score 6 but not double 3',	$E_7 \equiv$ 'score 7',
$E_8 \equiv$ 'score 8 but not double 4',	$E_9 \equiv$ 'score 9',
$E_{10} \equiv$ 'score 10 but not double 5',	$E_{11} \equiv$ 'score 11',

We will then have

$$P(W|H) = \sum_{i=2}^{11} P(E_i|H) P(W|E_i \wedge H) \, , \tag{3.2}$$

but each of these component probabilities requires further elaboration.

Consider for example $P(E_i|H)$. Elaborating as in the example at the end of Chapter 2, Section 2.6, we find the following probabilities.

$$P(E_2|H) = {}^3/_{18}, \qquad P(E_3|H) = {}^1/_{18},$$
$$P(E_4|H) = {}^1/_{18}, \qquad P(E_5|H) = {}^2/_{18},$$
$$P(E_6|H) = {}^2/_{18}, \qquad P(E_7|H) = {}^3/_{18},$$
$$P(E_8|H) = {}^2/_{18}, \qquad P(E_9|H) = {}^2/_{18},$$
$$P(E_{10}|H) = {}^1/_{18}, \qquad P(E_{11}|H) = {}^1/_{18}. \tag{3.3}$$

We now require the probabilities $P(W|E_i \wedge H)$. Some of these are easy, for if the Banker rolls '7, 11 or doubles' the Player will certainly not win. Thus

$$P(W|E_2 \wedge H) = 0, \qquad P(W|E_7 \wedge H) = 0,$$
$$P(W|E_{11} \wedge H) = 0. \tag{3.4}$$

The others are more difficult. We will elaborate them by extending the argument further to include the result of the Player's first roll. Consider for instance $P(W|E_6 \wedge H)$ and define the following simple logical partition for the Player's first roll.

$F_1 \equiv$ '7, 11 or doubles',
$F_2 \equiv$ 'score 6 but not double 3',
$F_3 \equiv$ 'some other result'.

Clearly, $P(W|E_6 \wedge F_1 \wedge H) = 1$ and $P(W|E_6 \wedge F_2 \wedge H) = 0$. If he rolls something else, he must roll again and the position is exactly as it was before his first roll, so $P(W|E_6 \wedge F_3 \wedge H) = P(W|E_6 \wedge H)$. We find probabilities like $P(F_1|E_6 \wedge H)$ just as we did for the probabilities (3.3). Thus

$$P(W|E_6 \wedge H) = P(F_1|E_6 \wedge H) P(W|E_6 \wedge F_1 \wedge H)$$
$$+ P(F_2|E_6 \wedge H) P(W|E_6 \wedge F_2 \wedge H)$$
$$+ P(F_3|E_6 \wedge H) P(W|E_6 \wedge F_3 \wedge H)$$
$$= ({}^7/_{18}) \times 1 + ({}^2/_{18}) \times 0 + ({}^9/_{18}) \times P(W|E_6 \wedge H). \tag{3.5}$$

We can easily solve the equation (3.5) to obtain $P(W|E_6 \wedge H) = {}^7/_9$. By similar means we acquire all the remaining probabilities that we need.

$$P(W|E_3 \wedge H) = {}^7/_8, \qquad P(W|E_4 \wedge H) = {}^7/_8,$$
$$P(W|E_5 \wedge H) = {}^7/_9, \qquad P(W|E_6 \wedge H) = {}^7/_9,$$
$$P(W|E_8 \wedge H) = {}^7/_9, \qquad P(W|E_9 \wedge H) = {}^7/_9,$$
$$P(W|E_{10} \wedge H) = {}^7/_8. \tag{3.6}$$

Inserting (3.6), (3.4) and (3.3) into (3.2) gives our final answer.

$$P(W|H) = ({}^3/_{18}) \times 0 + ({}^1/_{18}) \times ({}^7/_8) + ({}^1/_{18}) \times ({}^7/_8)$$
$$+ \ldots + ({}^1/_{18}) \times ({}^7/_8) + ({}^1/_{18}) \times 0$$

$$= 0.4915 \,.$$

Therefore, the Player's probability of winning is very nearly 0.5 and, since the game is played as the basis of a bet at odds of 1, it is almost fair. It favours the Banker very slightly.

..........}

We conclude this section with a simple corollary of extending the argument.

The irrelevant information theorem

Let $E_1, E_2, \ldots, E_n$ be a partition given H. Let $P(F | E_i \wedge H) = c$, the same value for each $i = 1, 2, \ldots, n$. Then $P(F | H) = c$.

{..........

Proof.

Extending the argument,

$$P(F | H) = \sum_{i=1}^{n} P(E_i | H) \cdot c = c \sum_{i=1}^{n} P(E_i | H) = c \,.$$

..........}

The point of this theorem is that if on learning which element of a partition is true, You will assign the same probability to F, regardless of which element is true, then that information is irrelevant. Therefore Your probability for F should have that same value whether You have that information or not.

Exercises 3(a)

1. A deck of cards is known to be incomplete, since it contains only 50 cards. Let $D_j \equiv$ 'it contains j Diamonds', $j = 11, 12, 13$. You assign probabilities $P_1(D_{11} | H) = 0.1$, $P_1(D_{12} | H) = 0.3$, $P_1(D_{13} | H) = 0.6$. The deck is shuffled and a card is drawn. Measure your probability for $C \equiv$ 'card is a Diamond'.

2. You are asked to buy a present for a child's birthday. Your information H leaves you uncertain about the following propositions – $S \equiv$ 'the child likes skipping', $D \equiv$ 'the child likes dolls', $G \equiv$ 'the child is a girl'. You measure the following: $P_a(S | G \wedge H) = 0.5$, $P_a(D | G \wedge H) = 0.6$, $P_a(S | (\neg G) \wedge H) = 0.3$, $P_a(D | (\neg G) \wedge H) = 0.2$, and $P_a(G | H) = 0.8$. You regard D and S as independent given $G \wedge H$, and also independent given $(\neg G) \wedge H$. Show that they are not independent given H. [This illustrates a general result.]

3. A gardener buys 10 packets of marigold seeds. The seeds are all mixed together. He takes a seed and plants it. Let $D \equiv$ 'the seed germinates and produces double flowers', and $V_i \equiv$ 'the seed is of variety i', ($i = 1, 2, \ldots, 6$). Using

the information in the table below, measure $P(D|H)$.

Variety	1	2	3	4	5	6
No. packets bought	3	1	1	2	1	2
$P(D|V_i \wedge H)$	0.9	0.4	0.3	0.1	0.2	0.1

3.2 Bayes' theorem

More important even than extending the argument is Bayes' theorem. In its simplest form it is just a direct consequence of the asymmetry of the Multiplication Law.

$$P(E \wedge F|H) = P(E|H)P(F|E \wedge H) = P(F|H)P(E|F \wedge H).$$

Therefore, if $P(F|H) \neq 0$,

$$P(E|F \wedge H) = \frac{P(E|H)P(F|E \wedge H)}{P(F|H)}. \tag{3.7}$$

Consider, for example, the student taking a multiple-choice examination. $C \equiv$'he chooses the right answer', $K \equiv$'he knows the right answer'. The probability $P(K|C \wedge H)$ is of interest; given that he has chosen the right answer, what is the probability that he knows it, rather than just guessed? Using the probability measurements of the last section,

$$P_1(K|C \wedge H) = \frac{P_1(K|H)P_1(C|K \wedge H)}{P_1(C|H)} = \frac{0.3 \times 0.95}{0.46} = 0.62.$$

In the previous section we found the denominator, $P_1(C|H)$, by extending the argument, and the numerator, $P_1(K|H)P_1(C|K \wedge H)$, was one of the terms in that extension. Bayes' theorem is typically used in this way.

Bayes' theorem
Let F be a proposition for which $P(F|H) \neq 0$. If $E_1, E_2, \ldots, E_n$ is a partition given H, then it is also a partition given $F \wedge H$, with probabilities given by

$$P(E_i|F \wedge H) = \frac{P(E_i|H)P(F|E_i \wedge H)}{\sum_{j=1}^{n} P(E_j|H)P(F|E_j \wedge H)}, \tag{3.8}$$

for $i = 1, 2, \ldots, n$.

{...........

Proof.

For any of the E_is we can use (3.7) to give

$$P(E_i|F \wedge H) = \frac{P(E_i|H) P(F|E_i \wedge H)}{P(F|H)} ,$$

and if we now elaborate $P(F|H)$ by extending the argument to include the partition $E_1, E_2, \ldots, E_n$ (given H) we obtain (3.8). To show that the E_is are a partition given $F \wedge H$, consider

$$P(E_i \wedge E_j|F \wedge H) = \frac{P(E_i \wedge E_j|H) P(F|E_i \wedge E_j \wedge H)}{P(F|H)} . \qquad (3.9)$$

For $i \neq j$, $P(E_i \wedge E_j|H) = 0$ because the E_is are a partition given H. Putting this value into (3.9) immediately shows that the E_is are mutually exclusive given $F \wedge H$. Finally, since the probabilities (3.8) sum to one over i, the E_is are also exhaustive given $F \wedge H$.

...........}

The examination example is the case $n=2$, using the logical partition $(K, \neg K)$.

Notice that in practical use of (3.8) the probabilities $P(F|E_i \wedge H)$ appear as components in the elaboration of the probabilities $P(E_i|F \wedge H)$. This elaboration is effective if the first set of probabilities can be measured directly more easily than the second set. Bayes' theorem is important because this is often the case. There may, for instance, be a natural time ordering, and it is generally easier to consider probabilities of later events conditional on earlier ones. Such is the case with the examination example. To contemplate the probability of the student knowing the right answer given that he has chosen the right answer means thinking backwards in time. It is easier and more accurate to elaborate this probability in terms of probabilities whose conditioning follows the natural time sequence. Often, it is helpful to consider such cases in terms of cause and effect: we wish to infer cause from effect, whereas it is easier to consider probabilities of observing the effect given the cause. In other contexts we think of the propositions E_i as a set of hypotheses and F as representing data concerning these hypotheses.

{...........

Example 1

Consider Example 1 of the last section. Suppose that You do not see the die being tossed but observe that the white ball is drawn from the bag. What is Your probability now for the propositions (hypotheses) S_i? The denominator of Bayes' theorem is (from the previous section) $P_1(\neg R|H) = 1 - P_1(R|H) = 0.266$. We find $P(S_1|H) = (^1/_6 \times ^1/_2)/0.266 = 0.313$, and similarly

$$P(S_2|H) = 0.209 , \qquad\qquad P(S_3|H) = 0.157 ,$$

$$P(S_4|H) = 0.126, \qquad\qquad P(S_5|H) = 0.105,$$
$$P(S_6|H) = 0.090.$$

..........}

{..........

Example 2

In Example 3 of the previous section consider $P(E_i|W \wedge H)$ for $i = 2, 3, \ldots, 11$. Using (3.8), the numerators are

$$P(E_2|H)P(W|E_2 \wedge H) = (^3/_{18}) \times 0,$$
$$P(E_3|H)P(W|E_3 \wedge H) = (^1/_{18}) \times (^7/_8),$$
$$P(E_4|H)P(W|E_4 \wedge H) = (^1/_{18}) \times (^7/_8),$$

....

$$P(E_{10}|H)P(W|E_{10} \wedge H) = (^1/_{18}) \times (^7/_8),$$
$$P(E_{11}|H)P(W|E_{11} \wedge H) = (^1/_{18}) \times 0, \qquad\qquad (3.10)$$

The denominator is the same in each case, i.e. the sum of these quantities. We have already computed the sum as $P(W|H) = 0.4915$. Dividing all the quantities (3.10) by this number gives

$$P(E_2|W \wedge H) = 0, \qquad\qquad P(E_3|W \wedge H) = 0.0989,$$
$$P(E_4|W \wedge H) = 0.0989, \qquad\qquad P(E_5|W \wedge H) = 0.1758,$$
$$P(E_6|W \wedge H) = 0.1758, \qquad\qquad P(E_7|W \wedge H) = 0,$$
$$P(E_8|W \wedge H) = 0.1758, \qquad\qquad P(E_9|W \wedge H) = 0.1758,$$
$$P(E_{10}|W \wedge H) = 0.0989, \qquad\qquad P(E_{11}|W \wedge H) = 0,$$

..........}

3.3 Learning from experience

Bayes' theorem can be considered as a description of how Your probabilities are changed in the light of experience . Given initial information H, if You then discover the truth of F, Your beliefs change; Your probability $P(E|H)$ no longer measures Your current degree of belief in E, and should be replaced for this purpose by $P(E|F \wedge H)$. Bayes' theorem expresses the relationship between these two probabilities as

$$P(E|F \wedge H) = \left(\frac{P(F|E \wedge H)}{P(F|H)}\right) P(E|H). \qquad\qquad (3.11)$$

The factor $P(F|E \wedge H)/P(F|H)$ is applied to update Your beliefs about E in the light of the new information F. In the examination example, given only information H the partition has two probabilities $P_1(K|H) = 0.3$ and $P_1(\neg K|H) = 0.7$, reflecting Your initial belief that the student is not likely to know the right answer. Observing that he chooses the right answer changes

Your probabilities to $P_1(K|C \wedge H) = 0.62$ and $P_1(\neg K|C \wedge H) = 0.38$. These are perfectly plausible; knowing C should increase Your probability for K. Bayes' theorem makes it clear why this should be so, for the proposition C is much more probable given K than given $\neg K$ (and H).

In general, (3.11) shows that learning the truth of F increases Your probability for E if F is itself sufficiently probable given E. However, 'sufficiently probable' is seen as relative to $P(F|H)$. The process by which new data update knowledge can be seen even more clearly if we work in terms of odds, rather than probabilities.

$$O(E|H) = \frac{1 - P(E|H)}{P(E|H)} = \frac{P(\neg E|H)}{P(E|H)}.$$

Applying this to (3.11) and a similar expression for $P(\neg E|F \wedge H)$ we find

$$O(E|F \wedge H) = \left(\frac{P(F|(\neg E) \wedge H)}{P(F|E \wedge H)} \right) O(E|H). \tag{3.12}$$

The term in parentheses is sometimes known as the *Bayes factor*, and the role of the information F is to determine this factor. If F is more probable given E than given $\neg E$ then Your odds against E are decreased by learning F, and so Your probability for E increases. A Bayes factor of 1 does not change Your beliefs, whereas a very high or very low Bayes factor will change them considerably.

Turning to Bayes' theorem, we can think of equation (3.8) operating as follows. Your initial probabilities $P(E_i|H)$ for the various hypotheses are multiplied by the quantities $P(F|E_i \wedge H)$. They are then divided by a common denominator, whose action is to ensure that they sum to one. For any specific hypothesis E_i, its probability is increased by the information F if $P(F|E_i \wedge H)$ is large *relative* to the values $P(F|E_j \wedge H)$ $(j \neq i)$ for the other hypotheses. In the context of Bayes' theorem, $P(F|E_i \wedge H)$ is known as the *likelihood of* E_i from data F.

Likelihood

The quantity $P(F|E_i \wedge H)$ appearing in Bayes' theorem, (3.8), is called the likelihood of E_i from data F. The information F modifies the probabilities of the E_is in proportion to their likelihoods. Equation (3.12) shows that $O(E_i|F \wedge H)$ equals $O(E_i|H)$ multiplied by the Bayes factor, which is the ratio of the likelihood of $\neg E_i$ to the likelihood of E_i (from data F).

We now consider sequential learning. Suppose that You are interested in the proposition E. Starting from information H, You learn first the truth of F and then of G. You could compute Your probability $P(E|F \wedge G \wedge H)$ by updating twice as follows.

$$P(E|F \wedge H) = \frac{P(F|E \wedge H) P(E|H)}{P(F|H)} , \qquad (3.13)$$

$$P(E|G \wedge (F \wedge H)) = \frac{P(G|E \wedge (F \wedge H)) P(E|F \wedge H)}{P(G|F \wedge H)} . \qquad (3.14)$$

[Notice that in the second updating You add G to the knowledge $F \wedge H$. This is a different updating from that which You would perform if You learnt G but had not previously learnt F, so that You were adding G to knowledge H.] Alternatively, You could update in a single step, adding information $(F \wedge G)$ to Your initial knowledge H:

$$P(E|(F \wedge G) \wedge H) = \frac{P((F \wedge G)|E \wedge H) P(E|H)}{P((F \wedge G)|H)} .$$

For true probabilities, both routes must lead to the same final probability.

The sequential learning theorem

For any propositions E, F and G, and initial information H,

$$P(E|F \wedge G \wedge H) = \frac{P(G|E \wedge F \wedge H) \left(\frac{P(F|E \wedge H) P(E|H)}{P(F|H)} \right)}{(G|F \wedge H)}$$
$$= \frac{P(F \wedge G|E \wedge H) P(E|H)}{P(F \wedge G|H)} .$$

{...........

Proof

The first expression arises immediately from substituting (3.13) into (3.14). This then equals

$$\frac{P(G|F \wedge (E \wedge H)) P(F|E \wedge H) P(E|H)}{P(G|F \wedge H) P(F|H)} .$$

Collecting the first two terms in the numerator by the Multiplication Law gives $P(F \wedge G|E \wedge H)$. Similarly the denominator reduces to $P(F \wedge G|H)$.
...........}

The theorem is useful in practice because the two updating methods yield two different elaborations of the same final result. For measuring probabilities, one may be better than the other.

The following terminology is commonly used when learning via Bayes' theorem.

Prior and posterior

$P(E|H)$ in (3.7), or the probabilities $P(E_i|H)$ in (3.8), are known
as prior probabilities. The corresponding updated probabilities,
$P(E|F \wedge H)$ or $P(E_i|F \wedge H)$, are known as posterior probabilities.
These terms, 'prior' and 'posterior', refer to Your state of
knowledge before and after learning the information F.

'Prior' and 'posterior' are *only* meaningful in relation to some specific information, which is particularly clear in the context of sequential updating. In equa-
tion (3.13) the probability $P(E|F \wedge H)$ is a posterior probability – relative to
information F. In equation (3.14) this same probability is used as a prior probability – relative to information G.

Sequential learning is very neatly expressible in terms of odds and Bayes factors. For,

$$\frac{P(F \wedge G | \neg E \wedge H)}{P(F \wedge G | E \wedge H)} = \frac{P(G | \neg E \wedge F \wedge H)}{P(G | E \wedge F \wedge H)} \times \frac{P(F | \neg E \wedge H)}{P(F | E \wedge H)} . \quad (3.15)$$

Thus the Bayes factor for the combined updating is the product of the Bayes
factors for the two separate updates.

Exercises 3(b)

1. A train travels between stations A and B. Let $L_A \equiv$ 'train is late arriving at A'
and define L_B similarly. You measure $P_1(L_A|H) = 3/8$, $P_1(L_B|H) = 5/8$,
$P_1(L_A \wedge L_B|H) = 1/4$. Elaborate (i) $P(L_B|L_A \wedge H)$, (ii) $P(L_B|(\neg L_A) \wedge H)$.

2. Ten percent of patients referred to a pathology clinic with suspected Krupps'
disease are actually suffering from the disease. Of these 10%, 6% have the
more severe 'lymphatic' form of the disease, and the other 4% have the 'non-
lymphatic' form. A new, quick test is available with the following characteristics: it gives a positive result with probability one for a patient with lymphatic
Krupps' disease, with probability 0.75 for a patient with non-lymphatic Krupps'
disease, and with probability 0.2 for a patient without Krupps' disease. Given
that the test gives a positive result for a certain patient, what is the probability
that he has Krupps' disease?

3. Three prisoners, Atkins, Brown and Carp, have applied for parole. It
becomes known that only one has been granted parole, but the Governor cannot
announce which until the end of the month. Atkins pleads with him for more
information. He agrees to tell Atkins the name of one applicant who has been
*un*successful, on the following understanding.

(a). If Atkins succeeds, he will name Brown or Carp, with equal probabilities.

(b). If Brown succeeds, he will name Carp.

(c). If Carp succeeds, he will name Brown.

The Governor believes that this procedure will give no useful information to Atkins. He names Brown.

On receiving this information, Atkins reasons that either he or Carp will be paroled, and therefore his probability of parole has increased from $1/3$ to $1/2$. Is this reasonable, or is the Governor's belief correct?

Atkins communicates the information to Carp, who also now believes that his chance of parole is $1/2$. Do you agree?

3.4 Zero probabilities in Bayes' theorem

Returning again to the examination example, suppose that some other person, with information H^*, wishes to measure his probability $P(K|C \wedge H^*)$. He agrees with Your measurements of probabilities of C given K or $\neg K$, i.e. $P_1(C|K \wedge H^*) = 0.95$, $P_1(C|\neg K \wedge H^*) = 0.25$. But he has an even lower opinion of the student, expressed in $P_1(K|H^*) = 0.1$. Using Bayes' theorem he obtains $P_1(K|C \wedge H^*) = 0.30$. The information C has changed his prior belief quite strongly but, unlike You, he does not give a posterior probability greater than 0.5 to K. He believes the student is more likely to have guessed, and would require stronger evidence than C to change this belief. In general, the more extreme Your prior belief is, the stronger the evidence that would be needed to dissuade You. Let us now take this argument to its limit.

Consider the simplest form of Bayes' theorem, equation (3.7), and let $P(E|H) = 0$. Then You will have $P(E|F \wedge H) = 0$ for any F. In other words, if You are initially absolutely convinced that the proposition E is false then nothing at all will persuade You otherwise. Alternatively, if You believe that E is certainly true then You will still believe this in the face of any evidence to the contrary. For

$$P(E|H) = 1 \quad \therefore P(\neg E|H) = 0 \quad \therefore P(\neg E|F \wedge H) = 0$$

$$\therefore P(E|F \wedge H) = 1 .$$

Now all this is perfectly reasonable. If You are already completely convinced of the truth (or falseness) of E then any other evidence which You might acquire is irrelevant. If, for instance, E is logically false then indeed $P(E|H) = 0$ for *any* information H. Unfortunately, with the exception of the implications of logic, we can never really be certain of anything. This means that Your *true* probability $P(E|H)$ will not be zero unless You are genuinely convinced that E is false, e.g. it is logically false. If $P(E|H)$ is very small but positive You may choose to measure it as zero, although this could be misleading. Furthermore, if $P(E|H)$ is a component of an elaboration of some other probability then for some elaborations $P_1(E|H)=0$ could be very accurate, but for others it may be highly inaccurate. Wherever practical, we should use small probabilities such as 0.01, or large probabilities like 0.99, in preference to the absolute limits of 0 and 1. In this way we leave open the possibility of acquiring sufficient evidence to dissuade us from our prior prejudices. It is worth noting that a common source of error in direct measurement is overconfidence. A proposition which is thought to be very probable/improbable is often given an unrealistically high/low probability measurement.

Suppose now that $P(F|E \wedge H)=0$. Then again $P(E|F \wedge H)=0$. What this means is that, given H, You regard the proposition F as impossible if E is true; therefore if F is known to be true E must be false. (In fact E and F are mutually exclusive given H.) The piece of evidence F is therefore conclusive. We can see this in Example 2 of Section 3.2. The same warning may be given as before, that we should assert $P(F|E \wedge H)=0$ only if E and F are *logically* mutually exclusive. For otherwise the conclusion is that E is now deemed to be impossible, and will remain so whatever further evidence may be gathered.

Now in both the previous cases we have said that, whether $P(E|H)=0$ or $P(F|E \wedge H)=0$, the conclusion is $P(E|F \wedge H)=0$. This is true, and Bayes' theorem itself is true, only on the understanding that $P(F|H) \neq 0$. We now consider what we can say about $P(E|F \wedge H)$ if $P(F|H)=0$. In a sense, it is not meaningful to say anything at all in this case because the state of information represented by $F \wedge H$ is itself impossible. However, we have remarked in Section 2.3 that there is a distinction between a proposition having zero probability and it being logically impossible. Therefore when, in Chapter 7, we encounter such propositions we shall find that it is possible to construct an appropriate form of Bayes' theorem. It is also interesting to consider how $P(F|H)=0$ might arise through assigning zero probabilities to other propositions, as above. From extending the argument we have

$$P(F|H) = P(E|H)P(F|E \wedge H) + P(\neg E|H)P(F|\neg E \wedge H). \quad (3.16)$$

So the denominator, $P(F|H)$, in Bayes' theorem can only be zero if the numerator is also zero. In such a case the theorem would anyway become unusable because in mathematics dividing zero by zero is generally ambiguous. Nevertheless, mathematics contains many instances where the ambiguity can be removed and a meaning attached to 'zero divided by zero' – differentiation, for example. It is in such a way that we reformulate Bayes' theorem in Chapter 7.

Looking again at (3.16), there are two ways in which $P(F|H)$ can be given a zero value. Most obviously, $P(F|H)=0$ if both $P(F|E \wedge H)$ and $P(F|\neg E \wedge H)$ are zero. We cannot have both $P(E|H)$ and $P(\neg E|H)$ zero, but we can have $P(E|H)=0$ and $P(F|\neg E \wedge H)=0$, or $P(\neg E|H)=0$ and $P(F|E \wedge H)=0$. In these cases F is impossible because it is mutually exclusive with another proposition ($\neg E$ or E) which is certainly true.

3.5 Example: disputed authorship

Between 1787 and 1788 a number of essays, known today as 'the Federalist Papers', were published in the state of New York. Historians long disputed the authorship of 12 of these papers, claimed by some to have been written by Alexander Hamilton and by others to be by James Madison. The dispute was largely resolved in favour of Madison by the publication in 1964 of '*Inference and Disputed Authorship: The Federalist*' by F. Mosteller and D. L. Wallace. (published by Addison-Wesley, Reading, Mass.). Mosteller and Wallace showed by studying a mass of data derived from papers already acknowledged to have been written by each author, that the statistical evidence was strongly in favour of Madison. We look at just one of the items of data which they used. In 47 out of 48 papers known to be by Hamilton, the word 'innovation' does not appear at all. Madison used 'innovation' rather more commonly; out of 50 papers known to be by him the word is used at least once in 16 but not in the other 34. Now suppose that we study one of the disputed papers and find that the word 'innovation' does appear. This fact will help in judging its authorship, seeming to suggest that it is by Madison. So let Your information be $H = W \wedge H_0$, where $W \equiv$ 'this paper contains the word innovation'. Let $M \equiv$ 'it was written by Madison', and Your target probability is $P(M|H)$. Now Mosteller and Wallace's analysis of undisputed papers suggests the measurements

$$P_f(W|(\neg M) \wedge H_0) = 1/48 , \quad P_f(W|M \wedge H_0) = 16/50 . \tag{3.17}$$

I have used a subscript f to denote that these measurements are based on Mosteller and Wallace's data on the *frequency* of usage of 'innovation' in undisputed works. Chapter 8 is devoted to a careful statistical analysis of frequencies, and to a justification of frequency measurements like (3.17). Even without such a justification, it is clear that they are intuitively reasonable. If only one out of Hamilton's 48 known papers uses 'innovation' then, knowing that some other paper was by Hamilton ($\neg M$) suggests a value of $1/48$ for the probability that this paper contains that word (W).

We have represented Your prior knowledge by H_0, but in this problem we must recognize that there will be differences in prior beliefs between the various protagonists. We consider three fairly representative viewpoints. Let H_m be the prior knowledge of someone who believes strongly in Madison's authorship, so that $P_1(M|H_m) = 0.9$, and H_h is the prior knowledge of a strong Hamilton advocate with $P_1(M|H_h) = 0.1$. Finally, a layman with prior knowledge H_l might reason that similar numbers of papers are known to be attributable to each author, and arrive at $P_1(M|H_l) = 0.5$. In each case, the prior knowledge includes the Mosteller and Wallace data, so that (3.17) applies for all three, substituting H_m, H_h or H_l for H_0. Let us apply Bayes' theorem to each of these beliefs.

$$P(M \mid W \wedge H_0)$$

$$= \frac{P(W \mid M \wedge H_0) P(M \mid H_0)}{P(W \mid M \wedge H_0) P(M \mid H_0) + P(W \mid (\neg M) \wedge H_0) P(\neg M \mid H_0)} .$$

Thus,

$$P_1(M \mid W \wedge H_m) = \frac{(^{16}/_{50})(^9/_{10})}{(^{16}/_{50})(^9/_{10}) + (^1/_{48})(^1/_{10})} = 0.9928 ,$$

$$P_1(M \mid W \wedge H_h) = \frac{(^{16}/_{50})(^1/_{10})}{(^{16}/_{50})(^1/_{10}) + (^1/_{48})(^9/_{10})} = 0.6305 ,$$

$$P_1(M \mid W \wedge H_l) = \frac{(^{16}/_{50})(^1/_2)}{(^{16}/_{50})(^1/_2) + (^1/_{48})(^1/_2)} = 0.9389 .$$

All three now have much greater degrees of belief in Madison's authorship of the paper in question than they had previously. Even the Hamilton advocate must admit that it was more probably written by Madison.

This analysis becomes much simpler in terms of odds and Bayes factor. All three subjects have the same Bayes factor against M, which from (3.17) is

$$(^1/_{48}) / (^{16}/_{50}) = 0.0651 ,$$

or about $^1/_{15}$. However, they have completely different prior odds.

$$O_1(M \mid H_m) = ^1/_9 , \quad O_1(M \mid H_h) = 9 , \quad O_1(M \mid H_l) = 1 .$$

The Bayes factor, being small, is strongly in favour of M, and is responsible for converting even high prior odds against M to low posterior odds.

$$O_1(M \mid W \wedge H_m) = 0.0072 , \quad O_1(M \mid W \wedge H_h) = 0.586 ,$$

$$O_1(M \mid W \wedge H_l) = 0.0651 .$$

Mosteller and Wallace used much more evidence than we have here, obtaining their final Bayes factor by multiplying together the successive factors as in (3.15). For every one of the twelve disputed papers they found very small Bayes factors, the largest of the twelve being about $1/80$, and the next largest about $1/800$. Data D which provided a Bayes factor of $1/80$ would convince even our 'Hamilton advocate':

$$O_1(M \mid D \wedge H_h) = (^1/_{80}) \times O_1(M \mid H_h) = ^9/_{80}$$

$$\therefore P_1(M \mid D \wedge H_h) = \frac{1}{1 + 9/80} = 0.8989 .$$

The most important feature of this example is that, after observing the truth of the proposition W, the probability measurements of the three protagonists were much closer together than they had been before. Equally, if You were asked to measure *Your* probability for M given information $W \wedge H_0$, then the accuracy of the measurements (3.17) could lead You to quite an accurate elaborated measurement of $P(M|W \wedge H_0)$ even if You could only measure $P(M|H_0)$ poorly. Suppose, for instance, that You felt that Your true probability $P(M|H_0)$ could be anywhere between 0.4 and 0.6. Applying calculations like those above would show that Your true posterior probability $P(M|W \wedge H_0)$ lies between 0.911 and 0.958, and is therefore much more accurately measured than $P(M|H_0)$.

4

Trials and deals

This chapter is concerned with sequences of partitions which arise in certain kinds of repeated experiments, such as in drawing balls successively from a bag. If each ball is replaced in the bag before the next is drawn, then the draws are essentially identical and independent. Partitions of this kind are known as trials and are the subject of the first half of this chapter. Notions of independence are important here, and so we begin by extending the Multiplication Law in Section 4.1, and extending the definition of independence in Section 4.2. Trials are defined in Section 4.3. An important problem in practice is to determine the probability of obtaining certain numbers of balls of specific colours from a series of draws. This problem is solved for balls of two colours in Section 4.5, and in Section 4.6 for any number of colours. Some preliminary theory on numbers of ways of arranging things comprises Section 4.4.

In Section 4.7 we begin to consider deals, which are analogous to drawing balls without replacement. The partitions in a deal are not independent, and their structure is more complex than trials. This structure is examined intuitively in Section 4.8, then in a formal, mathematical way in Section 4.9. Section 4.10 obtains probabilities for drawing balls of given colours in a series of draws, and the relationship between deals and trials is considered in Section 4.11.

4.1 The product theorem

Throughout this chapter we will need to consider repeated conjunctions like $\bigwedge_{i=1}^{k} E_i$. We therefore begin by generalizing the Multiplication Law.

The product theorem
For any n propositions $E_1, E_2, \ldots, E_n$ and information H,

$$P(\bigwedge_{i=1}^{n} E_i | H) = \prod_{i=1}^{n} P(E_i | H \wedge \bigwedge_{j=1}^{i-1} E_j). \tag{4.1}$$

{...........

Proof
By mathematical induction. The theorem is trivially true for $n=1$, and also true by the Multiplication Law for $n=2$. Assume it true for $1 \leq n \leq k$ and let

$$G \equiv \bigwedge_{i=1}^{k} E_i, \qquad F \equiv \bigwedge_{i=1}^{k+1} E_i = G \wedge E_{k+1}.$$

By the Multiplication Law,

$$P(F|H) = P(G|H)P(E_{k+1}|H \wedge G).$$

Since the theorem is assumed true for $n=k$, i.e. for the proposition G,

$$P(F|H) = \left\{ \prod_{i=1}^{k} P(E_i|H \wedge \wedge_{j=1}^{i-1} E_j) \right\} P(E_{k+1}|H \wedge \wedge_{j=1}^{k} E_j)$$

$$= \prod_{i=1}^{k+1} P(E_i|H \wedge \wedge_{j=1}^{i-1} E_j).$$

...........}

Notice that the expression (4.1) depends on the order in which we arrange the propositions. Rearranging the sequence and applying the theorem again gives a different but equally valid result. For two propositions we have

$$P(E \wedge F|H) = P(E|H)P(F|E \wedge H) = P(F|H)P(E|F \wedge H).$$

The equality of these two products is the starting point for Bayes' theorem. For three propositions there are six different forms,

$$P(E \wedge F \wedge G|H) = P(E|H)P(F|E \wedge H)P(G|E \wedge F \wedge H)$$

$$= P(F|H)P(G|F \wedge H)P(E|F \wedge G \wedge H)$$

$$= P(G|H)P(F|G \wedge H)P(E|G \wedge F \wedge H)$$

$$= \cdots$$

4.2 Mutual independence

In Section 1.7 we introduced the notion of independence between two propositions by saying that if You learnt whether one of them was true it would not affect Your probability for the other. It is easy to imagine more than two propositions for which this property holds. In fact our earliest discussions about uncertainty, in Section 1.1, present us with a nice example. Consider eight propositions;

$E_1 \equiv$ 'the British prime minister weighs less than 110 pounds',

$E_2 \equiv$ 'the next president of the United States will be a woman',

$E_3 \equiv$ 'Isaac Newton was hit on the head by an apple',

$E_4 \equiv$ 'five countries have nuclear weapons',

$E_5 \equiv$ 'bank interest rates will fall next month',

$E_6 \equiv$ 'it rained in Moscow yesterday',

$E_7 \equiv$ 'the largest prime number less than 10^{10} exceeds 999999980',

$E_8 \equiv$ 'cigarette smoking causes lung cancer'.

These are quite unconnected propositions. If I were to learn, for instance, that E_2 and E_4 are true but E_3 is false (i.e. $E_2 \wedge (\neg E_3) \wedge E_4$ is true), then this knowledge would not affect my probabilities for any of the other five propositions. In this section we study this generalization of the notion of independence, from 2 to n propositions, which we shall call *mutual independence*.

{...........

Pairwise independence

The formal definition of mutual independence is rather complicated and first it may be helpful to see why the complication is necessary, and also why we use the term 'mutual independence' rather than just 'independence'. Here is an example with three propositions. Take the following cards from a standard deck;

Ace of Spades, Ace of Diamonds,

King of Hearts, King of Clubs.

Together they make a little deck of four cards. Shuffle it thoroughly and select one. Each card has a probability $1/4$ of being chosen. Now let $E_1 \equiv$ 'Ace' $E_2 \equiv$ 'Spades or Hearts' and $E_3 \equiv$ 'Hearts or Diamonds'. By trivial application of the symmetry probability theorem,

$$P(E_1|H) = P(E_2|H) = P(E_3|H) = 1/2,$$

$$P(E_1 \wedge E_2|H) = P(E_1 \wedge E_3|H) = P(E_2 \wedge E_3|H) = 1/4.$$

Since $P(E_1 \wedge E_2|H) = P(E_1|H) P(E_2|H)$, E_1 and E_2 are independent given H. Similarly, the other two pairs of propositions are independent given H. If You learn whether any one of these propositions is true then Your probability for each of the other two is unaltered. However, as soon as You learn the truth or falseness of any two of them, You will know for certain whether the third is true. These propositions are pairwise independent but not mutually independent. This is a contrived example, and in practice full mutual independence is a far more natural occurrence, but it needs a careful definition.

..........}

In defining mutual independence we want to say things like

$$P(E_3|(\neg E_1) \wedge E_2 \wedge H) = P(E_3|H),$$

i.e.

$$P((\neg E_1) \wedge E_2 \wedge E_3|H) = P((\neg E_1) \wedge E_2|H) P(E_3|H)$$

$$= P(\neg E_1|H) P(E_2|H) P(E_3|H). \qquad (4.2)$$

Now let

$$E_{i,0} \equiv \neg E_i, \quad E_{i,1} \equiv E_i \qquad (4.3)$$

for $i=1,2,3$. For instance, if $j_1=0$, $j_2=1$, $j_3=1$, then equation (4.2) could be written

$$P(E_{1,j_1} \wedge E_{2,j_2} \wedge E_{3,j_3} | H) \tag{4.4}$$

$$= P(E_{1,j_1} | H) P(E_{2,j_2} | H) P(E_{3,j_3} | H).$$

Our definition will require that (4.4) hold not just for $(j_1, j_2, j_3) = (0, 1, 1)$ but also for the other seven triplets of zeroes and ones – (0, 0, 0), (0, 0, 1), (0, 1, 0), (1, 0, 0), (1, 0, 1), (1, 1, 0) and (1, 1, 1). (The $2^3 = 8$ propositions in question form a logical partition.)

Mutual independence

The n propositions $E_1, E_2, \ldots, E_n$ are said to be mutually independent given H if

$$P(\wedge_{i=1}^n E_{i,j_i} | H) = \prod_{i=1}^n P(E_{i,j_i} | H) \tag{4.5}$$

for any of the 2^n possible combinations of $j_i = 0$ or 1, where $E_{i,0}$ and $E_{i,1}$ are defined by (4.3), for $i = 1, 2, \ldots, n$.

In our preliminary discussion of mutual independence we said that information about some of the E_is would not alter Your probabilities concerning the others. In fact this verbal definition corresponds precisely with the mathematical definition above. Certainly the verbal definition implies (4.5), for if we apply the product theorem to the left-hand side of (4.5), then every term of the product simplifies from

$$P(E_{i,j_i} | H \wedge \wedge_{k=1}^{i-1} E_{k,j_k})$$

to $P(E_{i,j_i} | H)$, as on the right-hand side. It takes a rather longer argument to show that (4.5) also implies our verbal definition. To start with, if the n E_is are mutually independent then any subset of m of these n are also mutually independent. To prove this note that for every $j_1, j_2, \ldots, j_{n-1}$

$$P(\wedge_{i=1}^{n-1} E_{i,j_i} | H)$$

$$= P(E_n \wedge \wedge_{i=1}^{n-1} E_{i,j_i} | H) + P((\neg E_n) \wedge \wedge_{i=1}^{n-1} E_{i,j_i} | H).$$

$$= P(E_n | H) \prod_{i=1}^{n-1} P(E_{i,j_i} | H) + P(\neg E_n | H) \prod_{i=1}^{n-1} P(E_{i,j_i} | H)$$

$$= \prod_{i=1}^{n-1} P(E_{i,j_i} | H).$$

Therefore the $n-1$ propositions $E_1, E_2, \ldots, E_{n-1}$ are mutually independent given H. The argument obviously works for any other $n-1$ of the propositions, so any subset of $n-1$ propositions are mutually independent given H. Repeating the argument proves that this is true of any $n-2$ propositions, and so on for any subset of $m < n$ propositions.

Therefore,

$$P(\wedge_{i=1}^{m} E_{i,j_i} | H \wedge \wedge_{i=m+1}^{n} E_{i,j_i}) = \frac{P(\wedge_{i=1}^{n} E_{i,j_i} | H)}{P(\wedge_{i=m+1}^{n} E_{i,j_i} | H)}$$

$$= \frac{\prod_{i=1}^{n} P(E_{i,j_i} | H)}{\prod_{i=m+1}^{n} P(E_{i,j_i} | H)}$$

$$= \prod_{i=1}^{m} P(E_{i,j_i} | H) = P(\wedge_{i=1}^{m} E_{i,j_i} | H).$$

Since the ordering of the E_i s is immaterial in (4.5), we see that the probability of any proposition concerning a subset of the E_i s is unaffected by knowledge concerning the remainder of the E_i s. This is precisely the meaning that we wish mutual independence to have.

Equation (4.5) is much neater than the product theorem (4.1), and so mutual independence is an important simplifying concept. In many problems we can identify groups of propositions which we regard as mutually independent. Nevertheless, however inevitable the judgement of independence may appear to be, as in the example of the eight propositions at the beginning of this section, there is no such thing as *logical* mutual independence. Mutual independence, like equi-probability, is always a personal judgement. Logical arguments only extend to matters of logical certainty and impossibility which, in terms of probability, translates into probabilities of one or zero. Only in the trivial and uninteresting case where every proposition is either logically true or logically false could we speak of 'logical' mutual independence.

Exercises 4(a)

1. A delivery of 12 lightbulbs contains 3 that are faulty, but it is not known which 3. What probability would you give to the proposition that none of the first 4 lightbulbs to be used from this delivery are faulty?

2. A bag contains one white ball and one red ball. Successive draws are made according to Polya's urn rule (Exercise 1(c)2). Prove that the (objective) probability that the first n balls drawn are all red is $1/(n+1)$.

3. Measure your probability for the proposition that on every one of the next five days the U. S. dollar will fall in value against the Deutschmark. (To be precise, define the dollar's value on each day to be its value on the New York stock exchange at the close of trading on that day.)

4. Verify the formula $P(F|H) = p^{10}$ used in Exercise 2(b)5, assuming that the E_i s are mutually independent given H.

5. The diagram shows part of a nuclear power plant's safety trip system. Overheating triggers a signal in the reactor R, which travels to the alarm bell B through the sections of cable $S_1, S_2, \ldots, S_8$. These cables can fail, for various reasons, and the signal will reach the bell if and only if there is a route from R to B which does not include any failed section. Let $F_i \equiv$ 'section S_i fails', $P(F_i|H) = a$ for $i = 1, 2, \ldots, 8$ and suppose that the F_i s are mutually independent given H. What is the probability that an overheating signal will successfully trigger the alarm bell? Calculate this probability when a equals 0.5, 0.1 and 0.01.

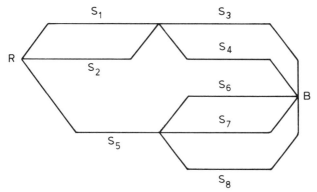

6. In Exercise 2(d)3, you might judge that knowing whether or not a person reads a daily newspaper would not affect your measurements of probabilities concerning the other three questions. Express this judgement formally as independence between two partitions. Are the measurements you made in Exercise 2(d)3 coherent with this independence judgement?

4.3 Trials

In this section we begin with a series of definitions. The next step up from independent propositions is independent partitions of propositions. Consider two logical partitions defined as follows. For $i = 1, 2, \ldots, 6$, let $E_i \equiv$ 'score i' when a die is tossed. For $j = 1, 2, \ldots, 5$, let $F_j \equiv$ 'the j-th candidate on the ballot paper will be elected' in an election with five candidates. Knowing which proposition in one partition is true does not affect my probabilities on the other partition. Again it is very easy to think of more examples like this, with two or more independent partitions.

Mutually independent partitions

For each $i=1, 2, \ldots, n$, let $E_{i,1}, E_{i,2}, \ldots, E_{i,m_i}$ be a logical partition of m_i propositions. These n partitions are said to be mutually independent given H if

$$P(\wedge_{i=1}^{n} E_{i,j_i} | H) = \prod_{i=1}^{n} P(E_{i,j_i} | H) \qquad (4.6)$$

for all $j_i = 1, 2, \ldots, m_i$ ($i = 1, 2, \ldots, n$).

Equations (4.6) and (4.5) are identical except for their circumstances. Mutual independence of propositions is just mutual independence of partitions, where each proposition E_i is formed into a partition of two by the addition of its negation $\neg E_i$. A partition of two propositions is called a binary partition. Section 4.5 concerns mutually independent binary partitions, and Section 4.6 generalizes the results to non-binary partitions. In both sections we shall be concerned with partitions which are also *similar*.

Similar partitions

For $i = 1, 2, \ldots, n$ let $E_{i,1}, E_{i,2}, \ldots, E_{i,m}$ be n logical partitions of m propositions each. They are said to be similar given H if

$$P(E_{1,j} | H) = P(E_{2,j} | H) = \cdots = P(E_{n,j} | H)$$

for $j = 1, 2, \ldots, m$.

You regard partitions as similar given H if You attach the same probabilities to corresponding components of the different partitions. Finally, we give a special name to a set of mutually independent, similar partitions.

Trials

A set of n logical partitions, each of m propositions, which are similar and mutually independent given H are said to comprise n m-way trials given H.

(When $m=2$ we say 'binary' instead of '2-way'.)

We have now rapidly introduced a number of concepts, and jargon to match. Actually, trials are very simple.

{...........
Examples

Two dice are to be tossed, one blue and one green. Let $E_{1,i} \equiv$ 'blue die shows i spots', and $E_{2,i} \equiv$ 'green die shows i spots'. The two partitions are independent and similar, so we have two 6-way trials.

A bag contains 12 balls, identical except for colour. One is black, two are green, three are red and six are yellow. A series of n draws is to be made. For the i-th draw let $E_{i,1} \equiv$ 'black', $E_{i,2} \equiv$ 'green', $E_{i,3} \equiv$ 'red' and $E_{i,4} \equiv$ 'yellow'. After each draw the chosen ball is replaced and the bag shaken again before the next draw. Thus

$$P(E_{i,1}|H) = 1/12, \quad P(E_{i,2}|H) = 1/6,$$

$$P(E_{i,3}|H) = 1/4, \quad P(E_{i,4}|H) = 1/2,$$

and these probabilities are unaffected by You knowing the outcomes of any other draws. Therefore we have n 4-way trials given H.
...........}

In general, whenever we have repetitions of some experiment providing mutually independent partitions, then the fact that the repetitions are under essentially identical conditions results in the natural judgement that the partitions are similar, and hence trials. Indeed, this is the context from which the term 'trials' derives. The primary concern with repeated experiments is in the probabilities of obtaining specified numbers of outcomes of various types. For instance, in the above experiment of drawing a ball from a bag, if $n=10$ draws are made what is Your probability for drawing a red ball 5 times and a yellow ball 5 times? Before answering this question we require some theory about arrangements of objects.

4.4 Factorials and combinations

The number of different ways in which one can arrange n different objects in a line is $n(n-1)(n-2)\cdots 3.2.1$. This is because of the following argument. There are n ways of choosing the first object in the line. Whatever that first object may be, there are $(n-1)$ to choose from for the second. There are therefore $n(n-1)$ ways of choosing the first two objects. Then the third may be any of the remaining $(n-2)$ objects, and so on until there is only one way of choosing the last object. Here are the $4 \times 3 \times 2 \times 1 = 24$ ways of arranging the first four letters of the alphabet.

abcd, abdc, acbd, acdb, adbc, adcb,
bacd, badc, bcad, bcda, bdac, bdca,
cabd, cadb, cbad, cbda, cdab, cdba,
dabc, dacb, dbac, dbca, dcab, dcba.

This number, $n(n-1)(n-2) \cdots 3.2.1$, is called '*n factorial*', and is denoted by $n!$. The first few factorials are

$$1! = 1, \quad 2! = 2, \quad 3! = 6, \quad 4! = 24, \quad 5! = 120, \quad 6! = 720, \quad 7! = 5040.$$

Clearly, $n!$ increases very rapidly thereafter as n increases. (10! is more than 3 million.) It is convenient to include the convention that $0! = 1$. This is sensible because

$$(n-1)! = n!/n \ ,$$

and so we can let $0! = 1!/1 = 1$.

Factorials

The number of ways of arranging n distinct objects is $n(n-1)(n-2) \cdots 3.2.1$, which is called '*n* factorial' and is denoted by $n!$. By convention, $0! = 1$.

Now suppose that we have to arrange n objects but they are not all different. If some objects are indistinguishable from one another then there will be less than $n!$ different arrangements. For instance, if I cannot distinguish between the letters b and d then for every one of the above 24 arrangements there is another which is indistinguishable from it by swapping the letters b and d over. There are therefore only $24/2 = 12$ different arrangements, i.e.

abcb, abbc, acbb, bacb, babc, bcab,

bcba, bbac, bbca, cabb, cbab, cbba.

In general, if r of the n objects are indistinguishable, then every one of the $n!$ arrangements becomes indistinguishable from $r!-1$ others, obtained by rearranging the indistinguishable objects, leaving $n!/r! = n(n-1)(n-2) \cdots (r+1)$ different arrangements. Next suppose that the other $(n-r)$ objects are indistinguishable from one another, so that there are only two different types of object. The $n!/r!$ arrangements must now be divided also by the $(n-r)!$ ways of arranging these $n-r$ indistinguishable objects. As a result, the number of different arrangements of n objects, r of which are of one type and the remaining $n-r$ of another type, is

$$\frac{n!}{r!(n-r)!} = \frac{n(n-1)(n-2) \cdots (r+1)}{(n-r)!}$$

$$= \frac{n(n-1)(n-2) \cdots (n-r+1)}{r!} \ .$$

For instance, we can arrange 3 letter a's and 2 letter b's in the following $5!/(3!2!) = 120/(6 \times 2) = 10$ different ways.

aaabb, aabab, aabba, abaab, ababa,

abbaa, baaab, baaba, babaa, bbaaa.

We have a special symbol for this expression,

$$\binom{n}{r} \equiv \frac{n!}{r!(n-r)!} \ .$$

This notation is now standard, but there is an earlier notation which is found in some textbooks, namely ^nC_r. We still read $\binom{n}{r}$ as 'n C r' ('en see are').

Finally, suppose that the n objects are of m types. There are r_1 objects of one type, r_2 of a second type, ... , and r_m objects of an m-th type. $\sum_{i=1}^{m} r_i = n$. Objects of the same type are indistinguishable. Following the argument we used in obtaining the formula for $\binom{n}{r}$, we find that the number of different arrangements is

$$\frac{n!}{r_1! r_2! \cdots r_m!} \ .$$

Our symbol for this expression is

$$\binom{n}{r_1, r_2, \ldots, r_m} \equiv \frac{n!}{\prod_{j=1}^{m} r_j!} \ .$$

We call this a multinomial coefficient. The binomial coefficient is the case $m=2$, and in that case we conventionally collapse the general notation $\binom{n}{r, n-r}$ to $\binom{n}{r}$.

Binomial and multinomial coefficients

The number of ways of arranging n objects of m different types, where there are r_i objects of type i ($i=1, 2, \ldots, m$, $\sum_{i=1}^{m} r_i = n$), is

$$\binom{n}{r_1, r_2, \ldots, r_m} \equiv \frac{n!}{\prod_{j=1}^{m} r_j!} \ .$$

This is known as a multinomial coefficient.

In the case $m=2$, where there are r objects of one type and $n-r$ of the other, we use the more compact symbol

$$\binom{n}{r} \equiv \frac{n!}{r!(n-r)!} \ ,$$

instead of $\binom{n}{r, n-r}$. This is called a binomial coefficient.

Exercises 4(b)

1. Calculate the values of $\binom{5}{k}$ for $k=0, 1, 2, 3, 4, 5$.

2. Calculate $\left(\begin{smallmatrix}10\\5,2,3\end{smallmatrix}\right)$.

3. In how many ways can you arrange 8 balls, 5 of which are red and 3 white?

4. Prove that for any $0 < k \leq n$, $\binom{n}{k-1} + \binom{n}{k} = \binom{n+1}{k}$.

5. Twelve cars, comprising 4 Fords, 3 Volkswagens, 3 Citroens and 2 Jaguars, drive into a parking area with 12 spaces. In how many ways can the cars be arranged (regarding cars by the same manufacturer as indistinguishable)?

6. Repeat Exercise 5 when the parking area has 16 spaces.

4.5 Binomial probabilities

The simplest kind of repeated experiment is where the trial has only two possible outcomes, like a coin toss, thereby generating binary partitions. It is conventional to call one outcome a 'success' and the other a 'failure'. In coin tossing, for example, we may choose to call Heads a success and Tails a failure. The main result of this section is a formula for the probability of r successes out of n repetitions of the experiment.

The binomial probability theorem

Let there be n binary trials given H. Let Your probability for success in each trial, given H, be p. Let S_r be the proposition that the number of successes in the n trials is r. Then

$$P(S_r|H) = \binom{n}{r} p^r (1-p)^{n-r} . \qquad (4.7)$$

{...........

Proof

Denote the partitions by $E_{i,1} \equiv$ 'success in trial i' and $E_{i,0} = \neg E_{i,1}$, for $i = 1, 2, \ldots, n$. Then

$$P(E_{i,1}|H) = p , \quad P(E_{i,0}|H) = 1-p . \qquad (4.8)$$

Let $F_\mathbf{j} \equiv \bigwedge_{i=1}^n E_{i,j_i}$, determined by the sequence $\mathbf{j} = \{j_1, j_2, \ldots, j_n\}$. Then because the partitions are mutually independent given H equation (4.6) applies for $P(F_\mathbf{j}|H)$. If $\mathbf{j}$ contains r 1s and $n-r$ 0s then

$$P(F_\mathbf{j}|H) = p^r (1-p)^{n-r} . \qquad (4.9)$$

If this is not immediately obvious, the following argument explains (4.9) more fully. We can write (4.8) as

$$P(E_{i,j}|H) = p^j (1-p)^{1-j}$$

for $j = 0$ or 1. Then from (4.6),

$$P(F_j|H) = \prod_{i=1}^{n} p^{j_i}(1-p)^{1-j_i} = p^r(1-p)^{n-r} , \qquad (4.10)$$

where $r = \sum_{i=1}^{n} j_i$ is the number of 1s in $\mathbf{j}$.

Now F_j is a proposition which asserts that there are r successes and $n-r$ failures in the n partitions, but it is not the same as proposition S_r. F_j says that r *specific* partitions give successes, whereas S_r simply asserts that the number of successes is r. There are perhaps many sequences $\mathbf{j}$ giving $\sum_{i=1}^{n} j_i = r$, so let these sequences be denoted by $\mathbf{j}_1, \mathbf{j}_2, \ldots, \mathbf{j}_t$. Then

$$S_r = \bigvee_{k=1}^{t} F_{\mathbf{j}_k} .$$

Since the $F_{\mathbf{j}_k}$s are logically mutually exclusive, and since every one of them has the same probability (4.10) given H, we have

$$P(S_r|H) = \sum_{k=1}^{t} P(F_{\mathbf{j}_k}|H) = t\, p^r(1-p)^{n-r} .$$

It remains to note that the number of sequences containing r 1s and $n-r$ 0s is just $t = \binom{n}{r}$.

...........}

We have already met, in Section 2.2, an application of this theorem. The question was asked, What is Your probability for obtaining five Heads in ten tosses of a coin? Your probability for Heads on any one toss is $1/2$, and these propositions are mutually independent. Therefore the conditions of the theorem hold with $n=10$, $r=5$ and $p=1/2$. Equation (2.4) computes the required probability:

$$P(S_5|H) = \binom{10}{5}(1/2)^5(1/2)^5 = 0.246 .$$

{...........

Example
A factory makes silicon chips for computers. Despite testing, some faulty chips leave the factory and are only found to be faulty after they have been assembled by the computer manufacturer. From experience (part of his knowledge H), the quality control manager of the chip-producing factory gives a probability $P_1(E|H) = 0.05$ for the proposition E that a given chip will be found to be faulty. A consignment of $n=50$ chips leaves the factory: what is the quality control manager's probability for $F \equiv$ 'this consignment contains fewer than 3 faulty chips'? Letting $E_{i,1} \equiv$ 'i-th chip is faulty' (which we will call a success!) then we have

$$P_1(E_{i,1}|H) = 0.05$$

for each i. Suppose that he judges these propositions to be mutually independent given H. Then, using the binomial probability theorem as an elaboration,

$$P_1(S_r|H) = \binom{50}{r} 0.05^r \, 0.95^{50-r} \, ,$$

where $S_r \equiv \text{'}r$ faulty chips'. We require the probability of $F = S_0 \vee S_1 \vee S_2$, so we can further elaborate by means of the sum theorem to give

$$P_1(F|H) = \binom{50}{0} 0.95^{50} + \binom{50}{1} 0.05^1 0.95^{49} + \binom{50}{2} 0.05^2 0.95^{48} \, .$$

We find $\binom{50}{0} = 1$, $\binom{50}{1} = 50$, $\binom{50}{2} = 1225$ and thus $P_1(F|H) = 0.5405$ to four decimal places.

...........}

In the exercises below, the conditions of the theorem, that successes are mutually independent with common probability p, are to be assumed wherever they seem appropriate. The validity of such assumptions is examined properly in Chapter 8.

Exercises 4(c)

1. Twelve dice are tossed. Call the proposition of an even score (2, 4 or 6) on a die a 'success'. What is the probability of (i) exactly 6 successes, (ii) less than 3 successes?

2. In any one-second interval, the probability that a radioactive substance will emit one or more gamma rays is 0.6. Events in successive time periods are mutually independent. Find the probability that, in 15 separate one-second intervals, gamma rays are observed in exactly 10 intervals.

3. During tests, a new drug is administered to 8 rats. Letting $R_j \equiv \text{'rat } j$ shows a positive response' you measure $P_1(R_j|H) = 0.75$ (for $j = 1, 2, \ldots, 8$). You regard the R_js as mutually independent given H. Measure your probability that 6 or more rats will show positive responses.

4. Prove that for any $n > 1$ and $0 \leq p \leq 1$

$$\sum_{r=0}^{n} \binom{n}{r} p^r (1-p)^{n-r} = 1 \, .$$

[Hint: use the partition theorem and the binomial probability theorem.] Hence show that $\sum_{r=0}^{n} \binom{n}{r} = 2^n$.

5. A computer generates random digits $0, 1, \ldots, 9$. Your probability for $E_{i,j} \equiv \text{'}i$-th digit generated is j' $(i = 1, 2, \cdots; j = 0, 1, 2, \ldots, 9)$ is 0.1, independently of other digits. What is your probability that in a sequence of 10 digits the digit 7 occurs 4 times?

4.6 Multinomial probabilities

We now extend the previous theorem from binary to m-way trials. For example, suppose that a die is to be tossed 20 times, and we wish to evaluate probabilities for propositions concerning the numbers of ones, twos, threes, fours, fives and sixes that will be observed. Some of these can be derived as binomial probabilities. For instance, Your probability for the proposition that exactly 5 sixes will be thrown is

$$\binom{20}{5} (1/6)^5 (5/6)^{15} = 0.1294 ,$$

because the natural 6-way partitions can be collapsed into binary partitions, with a throw of six being identified as a success and anything else as a failure. However, for more complex propositions we require a more general theorem.

The multinomial probability theorem

Suppose that we have n m-way trials given H in which, for $j=1, 2, \ldots, m$, the probability of the proposition of type j is

$$P(E_{i,j}|H) = p_j$$

for $i=1, 2, \ldots, n$. Let $S(r_1, r_2, \ldots, r_m) \equiv$ 'in the n trials r_1 propositions of type 1, r_2 of type 2, $\ldots, r_m$ of type m are true', where $\sum_{j=1}^{m} r_j = n$. Then

$$P(S(r_1, r_2, \ldots, r_m)|H)$$

$$= \binom{n}{r_1, r_2, \ldots, r_m} \prod_{j=1}^{m} p_j^{r_j} . \tag{4.11}$$

{..........

Proof.
Let $F_{\mathbf{j}} \equiv \bigwedge_{i=1}^{n} E_{i,j_i}$, determined by the sequence $\mathbf{j} = \{j_1, j_2, \ldots, j_m\}$ where each j_i is a number between 1 and m. Then for any $\mathbf{j}$ such that r_1 of the j_is equal 1, r_2 equal 2, ... , and r_m equal m we have

$$P(F_{\mathbf{j}}|H) = \prod_{i=1}^{n} P(E_{i,j_i}|H) = \prod_{i=1}^{n} p_{j_i} = \prod_{j=1}^{m} p_j^{r_j} . \tag{4.12}$$

The event $S(r_1, r_2, \ldots, r_m)$ states that one of $\binom{n}{r_1, r_2, \ldots, r_m}$ mutually exclusive $F_{\mathbf{j}}$s, each having probability (4.12), is true.
..........}

Thus, if 20 dice are tossed, Your probability of obtaining precisely 3 ones, 4 twos, 3 threes, 3 fours, 4 fives and 3 sixes is

$$\binom{20}{3,4,3,3,4,3}(^1\!/6)^3\,(^1\!/6)^4\,(^1\!/6)^3\,(^1\!/6)^3\,(^1\!/6)^4\,(^1\!/6)^3$$

$$= \frac{20!}{6^4\,24^2}\,\frac{1}{6^{20}} = 0.00000637\,.$$

{...........

Example
A certain breakfast cereal has a picture card in each pack. There are three
different pictures to collect. You wish to evaluate Your probability for the
proposition C that, if You buy four packs of cereal You will obtain a set of
three different pictures amongst the four cards. We identify four logical partitions, each of three propositions: $E_{i,j} \equiv$ 'card i has a picture of type j',
$(i=1,2,3,4;\ j=1,2,3)$. You have no reason to suppose that any of the three
pictures is more common than any other, and so You judge each partition to
be equi-probable given H , and therefore they are similar with

$$P_e\,(E_{i,j}|H) = ^1\!/3 \tag{4.13}$$

for all i and j. Finally, You judge that the partitions are mutually independent, so that they comprise four 3-way trials given H . The proposition C
can be written

$$C = S\,(2,1,1)\vee S\,(1,2,1)\vee S\,(1,1,2)\,.$$

Because of the equi-probability (4.13), we find that each of the three component propositions above has the same probability, and thence that

$$P_e\,(C|H) = 3\binom{4}{2,1,1}(^1\!/3)^4 = ^4\!/9\,.$$
...........}

Exercises 4(d)

1. In a certain population, the proportions of people having blood of types O, A,
B and AB are 42%, 28%, 18% and 12% respectively. If blood samples are
taken from four people, what probability would you give to the proposition that
they are all of different types?

2. A die has 1 spot on one face, 2 spots on two faces and 3 spots on three faces.
It is tossed 8 times. What is the probability that the total score equals 20?

3. A bag contains 10 balls – 4 red, 4 white and 2 green. Five balls are drawn,
with each ball being replaced before the next draw. What is the most probable
outcome, and what is its probability?

4. Four random digits are generated as in Exercise 4(c)5. What is your probability for $D \equiv$ 'more 7s than 2s'?

4.7 Deals

Trials have to be independent. In the remainder of this chapter we study partitions that have a particular kind of dependence. Consider dealing cards successively from a deck. The first card yields an equi-probable partition of 52 propositions. For instance, if $E_{1,1} \equiv$ 'card 1 is ace of spades' then

$$P(E_{1,1}|H) = 1/52 .$$

Now the second card is dealt. The first is not replaced in the deck. We can construct a second partition $E_{2,j}$, also of 52 propositions, but the two partitions are not independent. For if $E_{2,1} \equiv$ 'card 2 is ace of spades' then

$$P(E_{2,1}|E_{1,1} \wedge H) = 0 .$$

Thus if, in addition to the initial information H, You know also that the first card was the ace of spades, all cards except the ace of spades are possible (and would clearly be judged equi-probable) on the second deal, but $E_{2,1}$ is now logically impossible. In general,

$$P(E_{2,j_2}|E_{1,j_1} \wedge H) = 1/51$$

for all j_1 and j_2 between 1 and 52 provided that $j_1 \neq j_2$. Otherwise,

$$P(E_{2,j_1}|E_{1,j_1} \wedge H) = 0 .$$

A less extreme dependence arises if we consider coarser partitions. Let us redefine $E_{i,1} \equiv$ 'card i is an ace', but not necessarily the ace of spades, $E_{i,0} \equiv \neg E_{i,1}$. Then for a deal of n cards we have n binary partitions. For the first card,

$$P(E_{1,1}|H) = 4/52 = 1/13 .$$

Although $E_{1,1}$ and $E_{2,1}$ are not mutually exclusive (since it is possible for both the first two cards to be aces), they are still not independent. If the first is an ace, then only 3 of the remaining 51 cards are aces. Therefore for the second card,

$$P(E_{2,1}|E_{1,1} \wedge H) = 3/51 = 1/17 ,$$

whereas if the first is not an ace we will have

$$P(E_{2,1}|E_{1,0} \wedge H) = 4/51 .$$

However we define our partitions, card dealing produces partitions which are not mutually independent. Instead we have a well-defined structure of dependences of each deal on the results of previous deals. More generally, we could 'deal' from any collection of objects. The crucial feature is sampling without replacement.

Basic features of a deal

In a deal, objects are successively drawn from some collection. During the deal objects already dealt are not replaced in the collection and therefore cannot be dealt again. All objects remaining in the collection at any one time are equally likely to be dealt next.

Taking balls from a bag illustrates the general format. Consider a bag containing N balls identical except for colour. There are m different colours, and there are R_i balls of the i-th colour (for $i=1, 2, \ldots, m$), $R_1 + R_2 + \cdots + R_m = N$. Balls are drawn without replacement. Your probability for drawing a ball of a given colour at a given time will depend on how many balls of each colour You have observed to have been drawn already. The appropriate probabilities are given in the following general definition.

Definition of a deal

Let the propositions $E_{i,j}$, for $i=1, 2, \ldots, n$ and $j=1, 2, \ldots, m$ comprise n logical partitions of m propositions each. Let

$$P(E_{1,j}|H) = R_j/N ,$$

where $R_1, R_2, \ldots, R_m$ are positive integers and

$$n \leq N = \sum_{j=1}^{m} R_j .$$

Let $S(s_1, s_2, \ldots, s_m)$ be a proposition that in the first

$$p = \sum_{j=1}^{m} s_j < n \tag{4.14}$$

partitions, a specific s_j propositions of type j (for $j=1, 2, \ldots, m$) were observed to be true. Then the n partitions are said to be a deal of n from $(R_1, R_2, \ldots, R_m)$ if

$$P(E_{p+1,j}|S(s_1, s_2, \ldots, s_m) \wedge H) = \frac{R_j - s_j}{N - p} \tag{4.15}$$

for $j=1, 2, \ldots, m$ (and for every $p, s_1, s_2, \ldots, s_m$ satisfying (4.14)).

We can use these conditional probabilities (4.15) to find probabilities of propositions like $S(r_1, r_2, \ldots, r_m)$. For example, consider $n=2$ binary partitions where $E_{i,1} \equiv$ 'card i is an ace'. For $r=0, 1, 2$, let $S_r \equiv$ 'r aces are dealt'. Then

$$P(S_0|H) = P(E_{1,0} \wedge E_{2,0}|H) = P(E_{1,0}|H)P(E_{2,0}|E_{1,0} \wedge H)$$

$$= \frac{48}{52} \times \frac{47}{51} = 0.8507 .$$

$$P(S_1|H) = P((E_{1,1} \wedge E_{2,0}) \vee (E_{1,0} \wedge E_{2,1})|H)$$

$$= P(E_{1,1} \wedge E_{2,0}|H) + P(E_{1,0} \wedge E_{2,1}|H)$$

$$= P(E_{1,1}|H)P(E_{2,0}|E_{1,1} \wedge H)$$

$$\quad + P(E_{1,0}|H)P(E_{2,1}|E_{1,0} \wedge H)$$

$$= \frac{4}{52} \times \frac{48}{51} + \frac{48}{52} \times \frac{4}{51} = 0.1448 .$$

$$P(S_2|H) = P(E_{1,1} \wedge E_{2,1}|H) = P(E_{1,1}|H)P(E_{2,1}|E_{1,1} \wedge H)$$

$$= \frac{4}{52} \times \frac{3}{51} = 0.0045 .$$

In Section 4.10 we will obtain the general form of these equations, but using a neater method exploiting certain properties of deals. So our next section considers various properties implied by the above definition. These results are first deduced heuristically, then proved formally in Section 4.9. In particular we shall show that the n partitions which comprise the deal are similar.

4.8 Probabilities from information

A bag contains 5 balls, 3 red and 2 white, from which we will draw 3 balls, without replacement. Let $R_i \equiv$'ball i is red', and $W_i \equiv \neg R_i =$'ball i is white'. With the usual assumption that Your information is not special, the partitions (R_i, W_i), $i=1, 2, 3$, form a deal of 3 from $(3, 2)$ given H. For instance, the probability of a red ball on the first draw is

$$P(R_1|H) = 3/5 .$$

What is the probability of a red ball on the second draw? The answer is not: 'It depends on what colour ball is taken out on the first draw'. It is true, of course, that

$$P(R_2|R_1 \wedge H) = 2/4 , \quad P(R_2|W_1 \wedge H) = 3/4 :$$

this is all part of the deal structure (4.15). But neither of these probabilities is what we require. Instead we wish to know $P(R_2|H)$. Imagine that the first ball has been drawn, its colour has not been revealed to You, but You require to measure Your probability for R_2. Since You do not know which ball was drawn first, it is possible for any one of the five balls to be drawn on the second draw. You have no information to suggest that any of these is more probable than any other, and so it is natural to assign

$$P(R_2|H) = 3/5 .$$

An identical argument applies to the third draw, thus $P(R_3|H) = 3/5$ also.

Some people find this reasoning hard to accept. After all, at the third draw only 3 balls remain in the bag, and so it is only possible to draw one of these three. The crucial fact is that You do not know which three remain. Consequently You cannot say that any specific ball, out of the original five, cannot be drawn. You have no reason to give a higher or lower probability to any specific ball than to any other. An interesting consequence of this argument is that the three partitions (R_i, W_i) are similar. The argument clearly holds quite generally, and the partitions comprising a draw are always similar.

Similarity of deals
The partitions comprising a deal are similar.

Now consider $P(W_1|R_2 \wedge H)$. Knowing that the second ball to be drawn is red means that that ball could not have been drawn on the first draw. We still do not know which red ball was drawn but this information would not have been useful anyway (the irrelevant information theorem): whichever red ball was drawn on the second draw, the first must have been one of 4 balls, 2 of which are known to be red and 2 white. Therefore,

$$P(W_1|R_2 \wedge H) = 2/4 = 1/2,$$

the same as $P(W_2|R_1 \wedge H)$, and You would give the same value also to $P(W_3|R_1 \wedge H)$, $P(W_2|R_3 \wedge H)$, and so on. Again, the reasoning applies to deals generally.

Information in deals
Knowing that certain types of objects have been, or will be, drawn in some of the partitions of a deal, Your probabilities for any other partitions are as if the appropriate numbers of objects of those types were not in the collection.

For example, in a deal of 6 from (4, 5), if You know that objects of type 1 were drawn in the first and fifth draws, then Your probability for an object of type 1 at any other draw becomes $2/7$.

The above results are important, but before proceeding further we must examine their status carefully. In the last section, we defined a deal in the most natural way. We picture a deal being carried out sequentially, and so we defined a sequence of probabilities, each conditional on what has gone before. Using the definition alone we could *derive* $P(R_2|H)$ by extending the argument to include the first draw.

$$P(R_2|H) = P(R_2|R_1 \wedge H)P(R_1|H) + P(R_2|W_1 \wedge H)P(W_1|H)$$

$$= (^2/_4)(^3/_5) + (^3/_4)(^2/_5) = {}^3/_5 .$$

However, we arrived at this value more directly, by a consideration simply of what information You had (or did not have). The above piece of algebra serves to demonstrate that the various probability judgements are coherent. It is always useful and reassuring to check for coherence in this way because, however natural and obvious one's judgements of probabilities may seem, intuition can be false. There are, for instance, several 'paradoxes' in probability which arise from unwary, non-coherent assertions of equi-probability or independence.

In the next section we prove quite generally that probabilities measured as in this section, i.e. by considering just what information You have, may also be derived formally from (and hence cohere with) the probabilities (4.15) which define a deal. The method is to show that the natural time-sequence of a deal is actually irrelevant: probabilities are unchanged if we take the partitions in any other order. The key to this proof lies in considering the various possible outcome sequences in a deal. In our example, for instance, the proposition that all three balls are red is $G_1 \equiv R_1 \wedge R_2 \wedge R_3$. There are six other possible sequences,

$$G_2 \equiv R_1 \wedge R_2 \wedge W_3 , \qquad G_3 \equiv R_1 \wedge W_2 \wedge R_3 ,$$

$$G_4 \equiv W_1 \wedge R_2 \wedge R_3 , \qquad G_5 \equiv R_1 \wedge W_2 \wedge W_3 ,$$

$$G_6 \equiv W_1 \wedge R_2 \wedge W_3 , \qquad G_7 \equiv W_1 \wedge W_2 \wedge R_3 .$$

The proposition $W_1 \wedge W_2 \wedge W_3$ is not a possibility because there are only two white balls in the bag, therefore $G_1, G_2, \ldots, G_7$ form a logical partition. From the definition of the deal we find

$$P(G_1|H) = (^3/_5)(^2/_4)(^1/_3) = {}^1/_{10} ,$$

$$P(G_2|H) = (^3/_5)(^2/_4)(^2/_3) = {}^1/_5 .$$

The G_ts are not an equi-probable partition. However, suppose that we now draw all five balls from the bag, resulting in a *complete deal* of 5 from (3, 2). There are now 10 possible outcome sequences:

$$F_1 \equiv R_1 \wedge R_2 \wedge R_3 \wedge W_4 \wedge W_5 , \quad F_2 \equiv R_1 \wedge R_2 \wedge W_3 \wedge R_4 \wedge W_5 .$$

$$F_3 \equiv R_1 \wedge R_2 \wedge W_3 \wedge W_4 \wedge R_5 , \quad F_4 \equiv R_1 \wedge W_2 \wedge R_3 \wedge R_4 \wedge W_5 ,$$

$$F_5 \equiv R_1 \wedge W_2 \wedge R_3 \wedge W_4 \wedge R_5 , \quad F_6 \equiv R_1 \wedge W_2 \wedge W_3 \wedge R_4 \wedge R_5 ,$$

$$F_7 \equiv W_1 \wedge R_2 \wedge R_3 \wedge R_4 \wedge W_5 , \quad F_8 \equiv W_1 \wedge R_2 \wedge R_3 \wedge W_4 \wedge R_5 ,$$

$$F_9 \equiv W_1 \wedge R_2 \wedge W_3 \wedge R_4 \wedge R_5 , \quad F_{10} \equiv W_1 \wedge W_2 \wedge R_3 \wedge R_4 \wedge R_5 ,$$

and if we now do the corresponding calculations we find that $P(F_t|H) = {}^1/_{10}$ for every $t = 1, 2, \ldots, 10$, i.e. the F_ts are an equi-probable partition. It turns out that the possible outcome sequences in a complete deal are always equi-probable, and this constitutes the first theorem which we prove in the next section.

This result also has an intuitive explanation. Imagine that a deck of cards is shuffled and then cards are dealt one by one from the top. Clearly the possible sequences are just the possible arrangements of the cards in the deck before the deal begins. It is natural to suppose (as a result of the shuffling, perhaps) that all these possible arrangements are equi-probable. In general, the process of randomly dealing a complete set of N objects is analogous to arranging them randomly, i.e. shuffling.

{ ....

Coherence checking

This last argument is a nice illustration of the danger of accepting judgements uncritically without checking them for coherence. For if we are not thinking carefully, we might suppose that the argument would apply also to incomplete deals, and hence that the outcome sequences of an incomplete deal should be equi-probable, too. As we have just seen (in the propositions $G_1, G_2, \ldots, G_7$) this is not true. Discovering a non-coherence like this causes us to consider our reasoning more carefully. We will then see that to talk of shuffling the objects in an incomplete deal is meaningless because different outcome sequences will in general contain different sets of n objects.

..........}

Exercises 4(e)

1. A bag contains 5 red balls, 2 green and 4 white. They are drawn successively without replacement. What is the probability that the first is red, the third is white and the sixth is green?

2. A poker player has four Hearts and one Spade. He discards the Spade, hoping to draw a Heart. He has already observed 3 Clubs, 1 Diamond and 1 Heart discarded by his opponents. What probability should he give to drawing another Heart?

3. A hockey match ended 3-2. Measure your probability that the first and last goals were both scored by the losers.

4.9 Properties of deals

In this section we prove that the intuitive probability judgements of Section 4.8 cohere with the definition of a deal in Section 4.7. Less mathematically inclined readers may skip to Section 4.10.

Consider first the outcome sequences which can arise in a complete deal of N from $(R_1, R_2, \ldots, R_m)$. As in the definition of a deal, let $E_{i,j} \equiv$ 'i-th object dealt is of type j' $(i=1, 2, \ldots, N; j=1, 2, \ldots, m)$. The $E_{i,j}$s form N partitions –

$(E_{i,1}, E_{i,2}, \ldots, E_{i,m})$ for $i = 1, 2, \ldots, N$. Each possible outcome sequence is identified with the sequence of types $\mathbf{j} = \{j_1, j_2, \ldots, j_N\}$, where each j_i is a number from 1 to m inclusive. Thus $F_{\mathbf{j}}$ is the proposition

$$F_{\mathbf{j}} \equiv \bigwedge_{i=1}^{N} E_{i,j_i} . \qquad (4.16)$$

The notation is so far identical to Section 4.5, but there is an important difference. Not all sequences $\mathbf{j}$ for which each j_i is between 1 and m are possible. In a complete deal of N from $(R_1, R_2, \ldots, R_m)$ we will necessarily observe precisely R_1 objects of type 1, R_2 of type 2, and so on. Therefore, the possible outcome sequences are identified with the sequences $\mathbf{j}$ that comprise all the

$$C \equiv \binom{N}{R_1, R_2, \ldots, R_m} \qquad (4.17)$$

possible ways of arranging N objects, R_j of which are of type j $(j = 1, 2, \ldots, m)$.

Theorem 1. The complete deal theorem.
In a complete deal of N from $(R_1, R_2, \ldots, R_m)$ given H, the propositions $F_{\mathbf{j}}$ defined by (4.16) form an equi-probable partition given H. Thus $P(F_{\mathbf{j}}|H) = 1/C$, where C is defined by (4.17).

{...........

Proof
Let $F_{\mathbf{j}_1}$ be the proposition that the first R_1 objects dealt are of type 1, the next R_2 are of type 2, $\ldots$, the last R_m are of type m. Thus

$$\mathbf{j}_1 = \{1, 1, \ldots, 1, 2, 2, \ldots, 2, 3, \ldots, m-1, m, m, \ldots, m\}$$

and

$$F_{\mathbf{j}_1} = E_{1,1} \wedge E_{2,1} \wedge \cdots \wedge E_{R_1,1} \wedge E_{R_1+1,2} \wedge E_{R_1+2,2}$$

$$\wedge \cdots \wedge E_{N-1,m} \wedge E_{N,m} .$$

Using the definition (4.15) repeatedly,

$$P(F_{\mathbf{j}_1}|H) = P(E_{1,1}|H) P(E_{2,1}|E_{1,1} \wedge H) \times \cdots$$

$$\times P(E_{R_1,1}|E_{1,1} \wedge E_{2,1} \wedge \cdots \wedge E_{R_1-1,1} \wedge H) \times \cdots$$

$$= \frac{R_1}{N} \frac{R_1-1}{N-1} \cdots \frac{1}{N-R_1+1} \times$$

$$\times \frac{R_2}{N-R_1} \frac{R_2-1}{N-R_1-1} \cdots \frac{1}{N-R_1-R_2+1} \times$$

$$\times \frac{R_3}{N-R_1-R_2} \cdots \frac{1}{R_m+1} \frac{R_m}{R_m} \frac{R_m-1}{R_m-1} \cdots \frac{1}{1}$$

$$= \frac{R_1! R_2! \cdots R_m!}{N!} = \frac{1}{C}.$$

Now consider what happens if we go through a calculation like this for some other F_j. The denominators will still go $N, N-1, N-2, \ldots, 1$ and so form $N!$. On the first i for which $j_i=1$ (i.e. the first time at which an object of type 1 is observed), the numerator will be R_1. The next i for which $j_1=1$ yields a numerator R_1-1, and so on. Therefore, although the numerators will come in a different sequence from the above, they will still form $R_1! R_2! \cdots R_m!$, and hence $P(F_j|H)=1/C$ for every possible $\mathbf{j}$.

...........}

One implication of the next theorem is the converse of the complete deal theorem. That is, starting with the assumption that the C outcome sequences F_j are an equi-probable partition we obtain the deal probabilities (4.15). We shall then have shown that the N partitions form a complete deal if and only if their outcome sequences are equi-probable. However, the theorem is more general and has a more important purpose. It shows that the sequence in which a deal is naturally defined is irrelevant, because the probabilities have the deal structure even if we change the order of the partitions.

Theorem 2

Let the C propositions, defined as in (4.16) and corresponding to the C possible outcome sequences for a complete deal of N from $(R_1, R_2, \ldots, R_m)$, be an equi-probable partition given H.

Let $S(s_1, s_2, \ldots, s_m)$ be a proposition that, in a specified $p=\sum_{j=1}^{m} s_j < N$ partitions, a specific $s_j \leq R_j$ propositions of type j $(j=1, 2, \ldots, m)$ are true. Let the i-th partition be *not* one of the p partitions referred to in $S(s_1, s_2, \ldots, s_m)$.

Then for all such $S(s_1, s_2, \ldots, s_m)$ and i, and for all $j=1, 2, \ldots, m$,

$$P(E_{i,j}|S(s_1, s_2, \ldots, s_m) \wedge H) = \frac{R_j - s_j}{N-p}.$$

{..........

Proof

We first obtain $P(S(s_1, s_2, \ldots, s_m)|H)$, using the symmetry probability theorem. Specifying the outcomes of these p partitions implies that in the remaining $N-p$ partitions we will necessarily observe $R_j - s_j$ propositions of type j, but any arrangement of these is possible. Therefore $S(s_1, s_2, \ldots, s_m)$

is the disjunction of

$$C_1 = \left(R_1 - s_1, R_2 - s_2, \ldots, R_m - s_m \right)^{N-p}$$

of the propositions F_j. Hence $P(S(s_1, s_2, \ldots, s_m)|H) = C_1/C$. Adding $E_{i,j}$ to $S(s_1, s_2, \ldots, s_m)$ leaves $N-p-1$ partitions whose outcomes will comprise $R_1 - s_1$ of type $1, \ldots, R_{j-1} - s_{j-1}$ of type $j-1, R_{j+1} - s_{j+1}$ of type $j+1$ $, \ldots, R_m - s_m$ of type m, but $R_j - s_j - 1$ of type j. Therefore,

$$P(E_{i,j} \wedge S(s_1, s_2, \ldots, s_m)|H) = C_2/C ,$$

where C_2 is the multinomial coefficient for these $N-p-1$ partitions. Finally, the probability we require is the ratio of these two,

$$P(E_{i,j}|S(s_1, s_2, \ldots, s_m) \wedge H) = C_2/C_1 .$$

Almost everything now cancels out, leaving just $R_j - s_j$ in the numerator and $N-p$ in the denominator.
............}

By making the p partitions referred to in $S(s_1, s_2, \ldots, s_m)$ the first p partitions, and letting $i = p+1$, this theorem becomes the converse of the complete deal theorem. But because it is true for any p partitions and any i, it implies that the probabilities are not affected by changing the order of the partitions.

Because of the complete deal theorem, theorem 2 is true if, instead of asserting the equi-probability of the F_js, its first sentence were the following.

'Let the propositions $E_{i,j}$ ($i=1, 2, \ldots, N$; $j=1, 2, \ldots, m$) form a complete deal of N from $(R_1, R_2, \ldots, R_m)$.'

Furthermore, we need not even assume a complete deal, as the next theorem shows. Nor do we need to specify the partitions and outcomes referred to in $S(s_1, s_1, \ldots, s_m)$.

Theorem 3. The exchangeability of deals theorem

Let the propositions $E_{i,j}$ ($i=1, 2, \ldots, n$; $j=1, 2, \ldots, m$) form a deal of n from $(R_1, R_2, \ldots, R_m)$. Let $N = \sum_{j=1}^{m} R_j \geq n$.

Let $S(s_1, s_2, \ldots, s_m)$ be a proposition that, in a (specified or unspecified) $p = \sum_{j=1}^{m} s_j < n$ partitions, a (specified or unspecified) $s_j \leq R_j$ propositions of type j ($j=1, 2, \ldots, m$) are true. Let the i-th partition be *not* one of the p partitions referred to in $S(s_1, s_2, \ldots, s_m)$.

Then for all such $S(s_1, s_2, \ldots, s_m)$ and i, and for all $j=1, 2, \ldots, m$,

$$P(E_{i,j}|S(s_1, s_2, \ldots, s_m) \wedge H) = \frac{R_j - s_j}{N - p} .$$

{...........

Proof

Theorems 1 and 2 cover the case $n = N$, i.e. a complete deal, for a completely specified $S(s_1, s_2, \ldots, s_m)$. First consider the effect of not specifying the partitions referred to in $S(s_1, s_2, \ldots, s_m)$. Suppose You know that in some p partitions s_j propositions of type j are true (for $j = 1, 2, \ldots, m$), but You do not know which propositions are true in which partitions. If You were given this extra information You could use Theorem 2 to give a probability of $(R_j - s_j)/(N - p)$ for $E_{i,j}$. The probability is the same whichever partitions are involved, and so, by the irrelevant information theorem, this same probability applies when the partitions are unspecified.

If the $E_{i,j}$s form an incomplete deal of $n < N$, we merely define a further $N - n$ partitions to complete the deal. These propositions may not be of direct interest to us, and it may seem artificial to introduce them, but they are simply a means of elaborating probabilities that are of interest. It is now clear that the theorem is true even for an incomplete deal.

...........}

We have now justified mathematically the intuitive arguments which were presented in the previous section. One final theorem presents our results in a different, and perhaps more appealing way.

Theorem 4. The conditional deals theorem

Consider any subset of n_1 partitions out of the n partitions comprising a deal of n from $(R_1, R_2, \ldots, R_m)$ given H. Let H_1 consist of information H plus the proposition that in n_0 other partitions (specified or unspecified), separate from the chosen subset of n_1 partitions $(n_1 + n_0 \le n)$, s_j^0 propositions of type j ($j = 1, 2, \ldots, m$) are true. $(\Sigma_{j=1}^m s_j^0 = n_0)$. Then the n_1 partitions comprise a deal of n_1 from $(R_1 - s_1^0, R_2 - s_2^0, \ldots, R_m - s_m^0)$ given H_1.

{...........

Proof

We have seen that the ordering of partitions in a deal is immaterial, so we choose a specific ordering for convenience. Let the n original partitions consist of propositions $E_{i,j}$ ($i = 1, 2, \ldots, n$ and $j = 1, 2, \ldots, m$) and let the chosen subset be the first n_1 partitions. In accordance with the definition of a deal, let $S(s_1, s_2, \ldots, s_m)$ be a proposition that in the first $p = \Sigma_{j=1}^m s_j \le n_1$ partitions, s_j of type j are true ($j = 1, 2, \ldots, m$). Let $s_j^1 = s_j + s_j^0$, ($j = 1, 2, \ldots, m$) and define $G(s_1^1, s_2^1, \ldots, s_m^1)$ to be a proposition that in the first p partitions, plus a further n_0 partitions (different from the first n_1) s_j^1

propositions of type j are true $(j=1, 2, \ldots, m)$. Then

$$S(s_1, s_2, \ldots, s_m) \wedge H_1 = G(s_1^1, s_2^1, \ldots, s_m^1) \wedge H ,$$

therefore from theorem 3 we have

$$P(E_{p+1,j} | S(s_1, s_2, \ldots, s_m) \wedge H) = \frac{R_j - s_j^1}{N - p - n_0}$$

$$= \frac{(R_j - s_j^0) - s_j}{(N - n_0) - p} . \tag{4.18}$$

Comparing (4.18) with (4.15), and noting that $N - n_0 = \sum_{j=1}^{m} (R_j - s_j^0)$, we see that the n_1 chosen partitions comprise a deal of n_1 from $(R_1 - s_1^0, R_2 - s_2^0, \ldots, R_m - s_m^0)$ given H_1.

..........}

4.10 Hypergeometric probabilities

We now derive a theorem which, in the case of n partitions forming a deal, is analogous to the multinomial probabilities theorem for n trials.

The hypergeometric probabilities theorem
Suppose that we have a deal of n from $(R_1, R_2, \ldots, R_m)$ given H. Let $S(r_1, r_2, \ldots, r_m) \equiv$ 'in the n partitions r_j propositions of type j are true' $(j=1, 2, \ldots, m)$, where $\sum_{j=1}^{m} r_j = n$. Then

$$P(S(r_1, r_2, \ldots, r_m) | H) =$$

$$\frac{\binom{n}{r_1, r_2, \ldots, r_m} \binom{N-n}{R_1 - r_1, R_2 - r_2, \ldots, R_m - r_m}}{\binom{N}{R_1, R_2, \ldots, R_m}} , \tag{4.19}$$

where $N = \sum_{j=1}^{m} R_j$.

{..........
Proof
Let

$$C \equiv \binom{N}{R_1, R_2, \ldots, R_m}, \quad C_1 \equiv \binom{n}{r_1, r_2, \ldots, r_m},$$

$$C_2 \equiv \binom{N-n}{R_1 - r_1, R_2 - r_2, \ldots, R_m - r_m} .$$

We first consider the following counting question: In how many ways can N objects, R_j of which are of type j $(j=1, 2, \ldots, m)$, be arranged in a line such that the first n objects in the line comprise r_j objects of type j $(j=1, 2, \ldots, m)$? Clearly, there are C_1 ways of arranging the first n, and every one of the arrangements we are trying to count must begin with one of these ways. Similarly, there are C_2 ways of arranging the remaining $N-n$, and every one of the arrangements we are trying to count must end in one of these ways. We can couple any of the C_1 beginnings with any of the C_2 endings, so that the number of different arrangements of all N is C_1C_2.

Returning to the theorem itself, consider the n partitions in question as the first n in a complete deal of N from $(R_1, R_2, \ldots, R_m)$ given H. Now all C arrangements for the complete deal are equi-probable, but we have just seen that C_1C_2 of these satisfy the proposition $S(r_1, r_2, \ldots, r_m)$. Therefore by the symmetry probability theorem $P(S(r_1, r_2, \ldots, R_m|H) = (C_1C_2)/C$.

..........}

In the simplest case, $m=2$, the *binary* hypergeometric probabilities are analogous to the binomial probabilities. For a deal of n from $(R, N-R)$ we have

$$P(S_r|H) = \frac{\binom{n}{r}\binom{N-n}{R-r}}{\binom{N}{R}}.$$

For example, suppose that a pool contains $N=20$ fish, $R=6$ of which are trout. An angler catches $n=5$ fish. What is the probability that he fails to catch any trout? It is natural to judge that the fisherman's catch forms a deal of 5 from (6, 14). We find (to 3 significant figures)

$$P(S_0|H) = \frac{\binom{5}{0}\binom{15}{6}}{\binom{20}{6}} = \frac{5005}{38760} = 0.129. \tag{4.20}$$

So the chance of not catching any trout is quite small. If we classified the 20 fish more thoroughly, into m species rather than trout/not-trout, we could use the general result (4.19) to find the probability of the angler's catch containing, say, 0 trout, 3 perch, 1 carp and 1 pike. Here is a more complex example.

{..........

Example
In an international athletics meeting between Norway, Sweden and Denmark each nation enters 6 athletes for the marathon. Under the (doubtful) assumption that they are all of comparable ability, the probability that the first 6 to finish will comprise 4 Swedes and 2 Norwegians is

$$\binom{6}{4,2,0}\binom{12}{2,4,6}/\binom{18}{6,6,6} = 0.0121.$$

..........}

Exercises 4(f)

1. Seven free tickets are available for a concert. Fifty students apply for the tickets, comprising 11 science, 24 arts and 15 social science students. The lucky 7 applicants are chosen from these 50 by a lottery. What probability would you give to the proposition that the winners comprise 2 science, 3 arts and 2 social science students?

2. The streets of a city are laid out in a rectangular grid, part of which is shown in the diagram. A woman walks from A to B, travelling only North or East. If she passes through the point C she will be robbed by a 'mugger'. Supposing that you judge all her routes from A to B to be equi-probable, what is the probability that she is robbed?

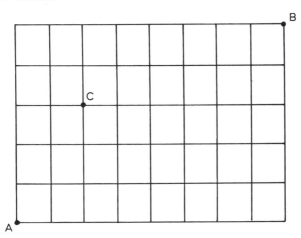

3. A regular deck of 52 cards is dealt to 4 bridge players, each receiving 13 cards. If you are one of these players, what is your probability of receiving exactly 5 hearts?

4.11 Deals from large collections

Consider again the angling example of the last section. An interesting thing happens if we increase the pool size N, keeping $n=5$ and the proportion of trout at 30%, i.e. $R = 0.3N$. First we increase the pool size to $N=50$, with $R=15$ trout. To distinguish this problem from the original one, we will let Your information this time be H_{50}, the subscript denoting the pool size. The original problem could be similarly identified by using H_{20}, so that equation (4.20) may be rewritten

$$P(S_0|H_{20}) = 0.129 . \tag{4.21}$$

We now compute the hypergeometric probability for S_0 given information H_{50} as

$$P(S_0|H_{50}) = \frac{\binom{5}{0}\binom{45}{15}}{\binom{50}{15}} = 0.153 . \tag{4.22}$$

Continuing to increase N we find

$$P(S_0|H_{200}) = 0.164 , \quad P(S_0|H_{1000}) = 0.167 ,$$

$$P(S_0|H_{4000}) = 0.168 . \tag{4.23}$$

Clearly, if the pool is large enough its precise size is unimportant, since the answer is more or less the same – 0.168 – for all large pools. To express this behaviour properly we require the mathematical notion of a limit.

{...........

Limits

We can express the behaviour (4.23) by saying that there is some quantity p such that as N increases $P(S_0|H_N)$ becomes closer and closer to p. For any $\varepsilon > 0$, however small, $|P(S_0|H_N) - p|$ (i.e. the absolute value of the difference) can be made less than ε simply by making N large enough. We say that $P(S_0|H_N)$ *tends to* p, or that p is the *limit* of $P(S_0|H_N)$, and we denote this by $P(S_0|H_N) \to p$. However, in doing this we need to specify the process under which the limit occurs. It happens as N increases without limit, which we denote by $N \to \infty$ and read as 'N tends to infinity'. Furthermore, R also increases without limit ($R \to \infty$) in such a way that the ratio R/N is fixed at 0.3.

...........}

The limiting behaviour we see in (4.21) to (4.23) can be observed in all hypergeometric probabilities. The following theorem identifies the limit probability.

Binomial limit theorem for hypergeometric probabilities

In a deal of N from $(R_1, R_2, \ldots, R_m)$ let each $R_j \to \infty$ such that their proportions

$$p_j = R_j/N , \tag{4.24}$$

where $N = \sum_{j=1}^{m} R_j$, remain constant. Then

$$P(S(r_1, r_2, \ldots, r_m)|H)$$

$$\to \binom{n}{r_1, r_2, \ldots, r_m} p_1^{r_1} p_2^{r_2} \cdots p_m^{r_m} . \tag{4.25}$$

{...........

Proof

We can write (4.19) as

$$P(S(r_1, r_2, \ldots, r_m)|H) = \left(r_1, r_2, \ldots, r_m\right)^n k$$

where

$$k = (N - n)! \frac{R_1!}{N!} \frac{R_1!}{(R_1-r_1)!} \frac{R_2!}{(R_2-r_2)!} \cdots \frac{R_m!}{(R_m-r_m)!}$$

$$= \frac{R_1}{N} \frac{R_1-1}{N-1} \cdots \frac{R_1-r_1+1}{N-r_1+1} \times$$

$$\times \frac{R_2}{N-r_1} \frac{R_2-1}{N-r_1-1} \cdots \frac{R_2-r_2+1}{N-r_1-r_2+1} \times$$

$$\times \frac{R_3}{N-r_1-r_2} \cdots \cdots \frac{R_3-r_3+1}{N-n+r_m+1} \times \cdots$$

$$\times \frac{R_m}{N-n+r_m} \frac{R_m-1}{N-n+r_m-1} \cdots \frac{R_m-r_m+1}{N-n+1} . \qquad (4.26)$$

Now each term in (4.26) is of the form $(R_j-a)/(N-b)$ which, as $N \to \infty$, tends to p_j.

..........}

The value of this result is practical as well as theoretical. For consider how equations (4.22) and (4.23) were calculated. For large R_js, the multinomial coefficients on the right hand side of (4.19) become enormous, and very laborious to calculate. Equation (4.26) is much more efficient, and for the angler's problem we simply have

$$P(S_0|H_N) = \frac{N-R}{N} \frac{N-R-1}{N-1} \frac{N-R-2}{N-2} \frac{N-R-3}{N-3} \frac{N-R-4}{N-4} .$$

This is how the calculations were performed. However, if n is large even (4.26) is impractical without a computer. The theorem tells us that whenever the R_js are all very large relative to the r_js, the hypergeometric probability (4.19) can be well approximated by the multinomial probability (4.25) (with p_js defined by (4.24)). For example, returning to the problem of the angler, we have $m=2$ and in each case $p_1=0.3, p_2=0.7$. The corresponding binomial probability is

$$P(S_0|H) = \binom{5}{0} (6/20)^0 (14/20)^5 = 0.7^5 = 0.168 .$$

For a sufficiently large pool, this is a good approximation to the correct hypergeometric probability, and is always simple to compute.

This close relationship between hypergeometric and multinomial probabilities is easy to explain in terms of a close relationship between deals and trials. In both cases we have n similar partitions, the difference being that trials are mutually independent whereas the partitions in a deal have dependencies expressed by the conditional probabilities (4.15). Yet these are also of the form $(R_j - a)/(N - b)$, and so as $N \to \infty$

$$P(E_{p+1,j} | S(s_1, s_2, \ldots, s_m) \wedge H) \to p_j .$$

But from the definition of a deal this equals $P(E_{1,j} | H)$, and since the ordering in a deal is irrelevant (and in particular its partitions are similar), it also equals $P(E_{p+1,j} | H)$. Therefore, the circumstances under which hypergeometric probabilities are approximated well by multinomial probabilities result in the n partitions comprising the deal becoming approximately independent, and thus in the deal itself being well approximated by a set of n trials. The approximation is also reasonable on an intuitive level: if the pool contains very many fish then information about a few of them does not change Your probability judgements for the others significantly.

Exercises 4(g)

1. A bag contains k red balls and $2k$ white balls. Five balls are drawn without replacement. Compute the probability of $S_2 \equiv$ '2 red and 3 white' for $k=2, 5, 10$. What is the limit of this probability as $k \to \infty$?

2. Seven students are to be chosen from the large number of students attending a certain college, where the proportions in the science, arts and social science faculties are 22%, 48% and 30% respectively. What is the probability of selecting 2 science, 3 arts and 2 social science students? Compare with Exercise 4(f)1.

5

Random variables

Many problems in probability measurement involve partitions like $T_r \equiv$ 'r tails in 10 tosses' ($r = 0, 1, \ldots, 10$), whose elements are associated with a number. It is convenient to be able to deal with the numbers themselves. A device which allows us to do this is the random variable. We can then operate easily with ideas such as the square of the number of tails in ten tosses, notions of an 'average' or 'expected' number of tails, and so on.

Section 5.1 defines a random variable and its probability distribution. Further definitions are introduced in Section 5.2, of joint, conditional and marginal distributions of two or more random variables. Versions of extending the argument and Bayes' theorem are derived in Section 5.3. Sections 5.4 and 5.5 give further extended examples of elaboration with random variables. In Section 5.6 notions of location and dispersion are introduced.

Specifying the probability distribution for a random variable can entail very many probability judgements. Sections 5.7 and 5.8 are concerned with more efficient ways of measuring distributions. The basic ideas are presented in Section 5.7, one of which is that a measured distribution is selected from a 'catalogue' of standard distributions. Section 5.8 defines various standard distributions.

5.1 Definitions

In elections for a city council, 50 councillors are to be elected. Consider the partition $C_0, C_1, \ldots, C_{50}$, where $C_t \equiv$ 't Conservative Party candidates are elected'. Uncertainty about the propositions in this partition is simply uncertainty about t, the number of Conservatives who are elected. In such a context, it is more natural to deal with this number directly, rather than through the partition $\{C_0, C_1, \ldots, C_{50}\}$. Formally, we define a *random variable* T which is said to take the *value* t if the proposition C_t is true. Other random variables could be defined from the same partition. For instance, if only two political parties are involved, we might define a random variable M whose value if C_t is true is $m = 2t - 50$, the margin for the Conservative Party over their opponents.

Random variables

A random variable X assigns a different value x_i to each element E_i of a partition. The proposition E_i is identified with the proposition that $X = x_i$, and for any H, $P(E_i | H) = P(X = x_i | H)$.

Thus if X is a random variable, expressions like '$X=1$', '$X>2$' or '$X^2+X-4=0$' are propositions. We may be interested, for instance, in a probability like $P((X>4)\vee(X<-1)|H)$. Using the sum theorem, any probability concerning X can be elaborated from the set of component probabilities $P(X=x_i|H)$, i.e. the probabilities on the underlying partition. These are the fundamental probabilities for a random variable, and are referred to collectively as its probability distribution.

Probability distribution
The probability distribution of a random variable X given information H is the set of probabilities $P(X=x|H)$ for all possible values x of X.

For brevity, we shall drop the qualifier 'probability' and refer to the *distribution* of X given H.

Many of the results of Chapter 4 can be translated into statements about the distributions of random variables.

{..........

Examples
The binomial probability theorem states that if R is the random variable which is the number of successes in n trials then its distribution given H is

$$P(R=r|H) = \binom{n}{r}p^r(1-p)^{n-r},$$

where p is the probability of success in a single trial given H. The possible values of R are $0, 1, \ldots, n$.

Let S be the number of propositions of type 1 which are true in a deal of m from $(r, n-r)$ given H. Then from the hypergeometric probabilities theorem its distribution given H is

$$P(S=s|H) = \binom{r}{s}\binom{n-r}{m-s}/\binom{n}{m}.$$

The possible values of S are not spelt out in that theorem. The minimum possible value is not necessarily zero, since if m exceeds $n-r$ then S must be at least $m-n+r$. A similar argument applies to the maximum value. The set of possible values for S ranges from $\max(0, m-n+r)$ to $\min(m, r)$.
..........}

A proposition based on one random variable may be expressed in terms of another random variable, and this may yield different elaborations of the target probability. Consider, for instance the proposition $F\equiv$'$(X-3)^2=4$'. Then $P(F|H)$ may be elaborated in terms of the distribution given H of the random variable X as

$$P(F \mid H) = P((X-3)^2 = 4 \mid H) = P(X=1 \mid H) + P(X=5 \mid H). \qquad (5.1)$$

Now consider the random variable $Y \equiv X-3$. This is defined on the same partition as X, but wherever X is given a value x, Y has the value $y = x-3$. The elaboration in terms of the distribution of Y is

$$P(F \mid H) = P(Y^2 = 4 \mid H) = P(Y=-2 \mid H) + P(Y=2 \mid H). \qquad (5.2)$$

Now (5.2) is actually the same elaboration as (5.1). Because the two random variables are defined on the same partition their distributions comprise the same probabilities; those probabilities are just attached to different values. However, if we now define the random variable $Z \equiv Y^2$, our target probability is simply

$$P(F \mid H) = P(Z=4 \mid H).$$

This is a different elaboration because Z is defined on a different partition. Remember that a random variable takes distinct values on the propositions in its underlying partition. Therefore if E_1 and E_5 are the propositions that '$X=1$' and '$X=5$', then E_1 and E_5 are both in the partition that underlies X (and Y), but $E_1 \vee E_5$ is in Z's partition.

5.2 Two or more random variables

In a great many problems we need to handle two or more random variables. First consider two random variables, X and Y. The probability given H of any proposition concerning these random variables can be elaborated in terms of the set of probabilities $P((X=x) \wedge (Y=y) \mid H)$, for all possible values x of X and y of Y. These probabilities comprise the *distribution of X and Y given H*. Sometimes, to emphasize that we are referring to a distribution of more than one random variable, we would call it the *joint* distribution of X and Y given H.

Through the Multiplication Law we have

$$P((X=x) \wedge (Y=y) \mid H) = P(X=x \mid H) P(Y=y \mid (X=x) \wedge H). \qquad (5.3)$$

which expresses the joint distribution as a product of two other distributions, one being a distribution of the single random variable X and the other of the single random variable Y. However, the two distributions are based on different information. The probabilities $P(X=x \mid H)$ form the distribution of X given H, which is based on the same information as the joint distribution. In this context we would refer to $P(X=x \mid H)$ as the *marginal* distribution of X. Since $P(Y=y \mid (X=x) \wedge H)$ is conditioned also on the value of X, we would in this context refer to it as the *conditional* distribution of Y given $X=x$ (and H).

Like the terms 'prior' and 'posterior', we can only use 'joint', 'marginal' and 'conditional' in a relative sense. Relative to any base information H, (5.3) says that the joint distribution of X and Y is the product of the marginal distribution of X with the conditional distributions of Y given X. (Notice that there is a different conditional distribution of Y for each possible value x of X.)

Two random variables
For two random variables X and Y, their (joint) distribution given H is the set of probabilities

$$P((X=x)\wedge(Y=y)|H),$$

for all possible values x of X and y of Y. Relative to any information H this joint distribution may be expressed as the product of the marginal distribution of X (i.e. $P(X=x|H)$) with the conditional distributions of Y given X (i.e. $P(Y=y|(X=x)\wedge H)$).

{...........

Example
In the card game 'bridge', a standard deck of 52 cards is dealt to the four players, so that each receives 13 cards. Suppose that You are one of the players and let A be (the random variable whose value is) the number of aces that You are dealt. The possible values of A are 0, 1, 2, 3, 4. We have a deal of 13 from (4, 48) and the distribution of A (as in the previous example) is

$$P(A=0|H) = \binom{13}{0}\binom{39}{4}/\binom{2}{4} = 0.304$$
$$P(A=1|H) = \binom{13}{1}\binom{39}{3}/\binom{52}{4} = 0.439$$
$$P(A=2|H) = \binom{13}{2}\binom{39}{2}/\binom{52}{4} = 0.213$$
$$P(A=3|H) = \binom{13}{3}\binom{39}{1}/\binom{52}{4} = 0.041$$
$$P(A=4|H) = \binom{13}{4}\binom{39}{0}/\binom{52}{4} = 0.003 \tag{5.4}$$

Now let B be the number of aces dealt to one of the other players. Given H and $A=a$, B's distribution is obtained from the resulting deal of 13 from $(4-a, 35+a)$. For each value of a we obtain a conditional distribution for B. For instance, the conditional distribution of B given H and $A=2$ is

$$P(B=0|(A=2)\wedge H) = \binom{13}{0}\binom{26}{2}/\binom{39}{2} = 0.439$$
$$P(B=1|(A=2)\wedge H) = \binom{13}{1}\binom{26}{1}/\binom{39}{2} = 0.458$$
$$P(B=2|(A=2)\wedge H) = \binom{13}{2}\binom{26}{0}/\binom{39}{2} = 0.105$$

The table below shows the complete set of conditional distributions. Each column shows the conditional distribution for a different value of A, and we can see that the numbers in each column sum to one.

a b	0	1	2	3	4
0	0.182	0.285	0.439	0.667	1.000
1	0.411	0.462	0.456	0.333	-
2	0.308	0.222	0.105	-	-
3	0.090	0.031	-	-	-
4	0.009	-	-	-	-

Table 5.1. $P(B=b \mid (A=a) \wedge H)$ in bridge game example

We can now use (5.3) to derive the joint distribution of A and B. This means multiplying the figures in the table above by the marginal distribution of A given in (5.4). The resulting table is presented below. Notice that this represents a single distribution, and therefore the sum of all the numbers in the table is one.

a b	0	1	2	3	4
0	0.055	0.125	0.094	0.027	0.003
1	0.125	0.203	0.097	0.014	-
2	0.094	0.097	0.022	-	-
3	0.027	0.014	-	-	-
4	0.003	-	-	-	-

Table 5.2. $P((A=a) \wedge (B=b) \mid H)$ in bridge game example.

To complete this example, we can confirm the above distribution directly from the hypergeometric probabilities theorem. To do this we have to imagine a different kind of deal. Consider a deck of 52 cards, 13 of which are labelled 'Me', 13 are labelled 'You' and the remaining 26 'Other'. Dealing four cards from this deck, and thinking of them as the four Aces, is an equivalent formulation of our problem. (Instead of dealing cards to people we are dealing people to cards.) The probability of a of these Aces being 'Me', b being 'You' and the remainder 'Other' is

$$P((A=a) \wedge (B=b) \mid H) = \binom{13}{a}\binom{13}{b}\binom{26}{4-a-b}/\binom{52}{4}.$$

It is easy to confirm that this formula yields Table 5.2.
............}

If the conditional distributions of a random variable X given $H \wedge (Y=y)$ are all the same as its marginal distribution then it is clear that learning the value of Y does not change Your beliefs about X. We say that X and Y are *independent* given H. It is permissible to use this symmetric terminology because we can easily show that the above situation will imply that learning the value of X will not change Your beliefs about Y. For we have

$$P(X=x|H\wedge(Y=y)) = P(X=x|H)$$

for all possible x and y. Multiplying both sides by $P(Y=y|H)$, and using the Multiplication Law gives

$$P((X=x)\wedge(Y=y)|H) = P(X=x|H)P(Y=y|H). \tag{5.5}$$

Then using the Multiplication Law as in (5.3) and dividing by $P(X=x|H)$ we have

$$P(Y=y|H\wedge(X=x)) = P(Y=y|H).$$

Since independence of random variables is symmetric, the symmetric expression (5.5) is preferred as a definition.

Independent random variables
Two random variables X and Y are said to be independent given H if

$$P((X=x)\wedge(Y=y)|H) = P(X=x|H)P(Y=y|H)$$

for all possible values x of X and y of Y.

If X and Y are independent given H then learning the value of either random variable (in addition to H) does not alter Your beliefs about the other.

For example, most people would judge that the random variables S, the height of the premier of the Soviet Union, and D, the distance from Addis Ababa to Zanzibar, are independent (given their current information). Knowledge of one cannot reasonably affect Your beliefs about the other.

If we have more than two random variables, there are many distributions which we can consider. For instance, suppose that we have defined five random variables named A, B, X, Y, Z. Then the probabilities

$$P((X=x)\wedge(Y=y)\wedge(Z=z)|(H\wedge(A=a))),$$

for all possible values x, y, z comprise the distribution of X, Y and Z given $H\wedge(A=a)$. This is a 'joint' distribution in the sense that it is a distribution of three random variables together. It is also a 'conditional' distribution in the sense that it is conditional on the value of A. Finally, it is 'marginal' in the sense that it does not involve, and is not conditional on, B. When dealing with three or more random variables, these terms are difficult to use unambiguously, and so we shall avoid them.

Nevertheless, all the other ideas which we have presented concerning two random variables generalize immediately to three or more. The joint distribution of several random variables may be decomposed via the product theorem, into a product of distributions for each of the random variables individually. The concept of independence generalizes to mutual independence.

Several random variables

The distribution given H of n random variables $X_1, X_2, \ldots, X_n$ comprises the probabilities

$$P(\wedge_{j=1}^{n}(X_j = x_j) | H)$$

for all possible values x_j of each X_j. It may be written as

$$P(\wedge_{j=1}^{n}(X_j = x_j) | H)$$

$$= \prod_{j=1}^{n} P(X_j = x_j | H \wedge \wedge_{k=1}^{j-1}(X_k = x_k)), \qquad (5.6)$$

where the terms on the right are distributions of individual random variables, each conditional on the values of all earlier random variables in the sequence.

The random variables are said to be mutually independent given H if

$$P(\wedge_{j=1}^{n}(X_j = x_j) | H) = \prod_{j=1}^{n} P(X_j = x_j | H). \qquad (5.7)$$

In this case, knowledge of any X_js (in addition to H) will not alter Your beliefs about any other X_js.

Exercises 5(a)

1. A farmer has 5 ewes, each of which will give birth to at least one lamb in the Spring. He will have more than 5 lambs if any of his ewes has twins. Let $E_i \equiv 'i$ ewes have twins' for $i = 0, 1, 2, 3, 4$ or 5. He assigns the following probabilities.

$$P(E_0|H) = 0.25, \quad P(E_1|H) = 0.32, \quad P(E_2|H) = 0.27,$$

$$P(E_3|H) = 0.13, \quad P(E_4|H) = 0.03, \quad P(E_5|H) = 0.$$

State his distributions, given H, for the following random variables: T = number of ewes with twins, L = number of lambs, S = number of single births, P = proportion of lambs which are twins.

2. Exercise 2(d)2 gives a (joint) distribution for the two random variables

$G=$number of goals for team X and $K=$number of goals for team Y, in a soccer match. Find the (marginal) distribution of G.

3. A company produces a new passenger aircraft. In the first six months after the aircraft is announced, let X be the number of orders received from U.S. airlines, and let Y be the number of orders from other airlines. Suppose that X and Y are independent given Your information H, with the following distributions.

n	0	1	2	3	4	5	6
$P(X=n|H)$	.40	.30	.25	.05	.00	.00	.00
$P(Y=n|H)$	.15	.20	.20	.15	.15	.10	.05

Write down Your joint distribution for X and Y given H. Letting $Z=X+Y$, the total number of orders, show that $P(Z=2|H)=0.1775$. Derive the remaining probabilities in Your distribution of Z given H.

5.3 Elaborations with random variables

Distributions of random variables are just collections of probabilities and so may be elaborated in the same ways as probabilities generally. Often, we can usefully apply essentially the same elaboration to each probability in a distribution. Equation (5.3) is an obvious example, where not only do the target probabilities on the left-hand side comprise a distribution but the component probabilities, seen on the right-hand side, may also be formed into distributions. The interpretation is that a joint distribution may be elaborated as the product of a marginal distribution and a set of conditional distributions.

Here is an example of the use of (5.3) as an elaboration. An employer receives six applications for a job vacancy, and is interested in whether the applications are from men or women. Let X be the number of women applicants in the first four applications to be opened, and let Y be the number of women in the remaining two applications. Their joint distribution comprises the 15 probabilities $P((X=x)\wedge(Y=y)|H)$ for $x=0, 1, 2, 3, 4$ and $y=0, 1, 2$. Suppose that the employer first measures her marginal distribution for X as in Table 5.3.

x	0	1	2	3	4
$P_e(X=x|H)$	0.25	0.25	0.20	0.15	0.15

Table 5.3 Distribution of X for the employment example

Now X and Y would not be independent. If $X=4$ then this suggests that the job is attractive to women, and hence that Y might equal two. Whereas if $X=0$ it suggests that Y will probably also be zero. This phenomenon is seen in the employer's conditional distributions for Y given X and H, tabulated below. The entries in Table 5.4 are $P(Y=y|H\wedge(X=x))$.

x	0	1	2	3	4
y					
0	0.7	0.6	0.45	0.3	0.1
1	0.2	0.25	0.3	0.35	0.3
2	0.1	0.15	0.25	0.35	0.6

Table 5.4. Conditional distributions of Y given X,
employment example

Using (5.3) we elaborate joint probabilities $P_e((X=x)\wedge(Y=y)|H)$ as follows.

x	0	1	2	3	4
y					
0	0.175	0.15	0.09	0.045	0.015
1	0.05	0.0625	0.04	0.0525	0.045
2	0.025	0.0375	0.05	0.0525	0.09

Table 5.5. Joint distribution, employment example

This is quite a good elaboration. It is difficult to measure the 15 joint probabilities directly so as to reflect properly the structure of the employer's beliefs.
That structure is captured in Table 5.4, where higher values of X are made to
correspond with a belief in higher values of Y. Another difficulty with measuring the joint distribution directly is that the individual probabilities are all rather
small. The component probabilities in the elaboration are generally larger,
which is an aid to more accurate measurement.

The two most important elaborations for distributions are analogues of the
two most important elaborations for individual probabilities – extending the
argument and Bayes' theorem.

Extending the argument (distributions)
For any two random variables X and Y, and information H,

$$P(X=x|H) = \sum_y P(X=x|(Y=y)\wedge H)P(Y=y|H),\qquad(5.8)$$

where the sum is over all possible values y of Y, and the result
applies for all possible values x of X.

Since the propositions '$Y=y$' for all possible y form a partition given H (from
the definition of a random variable), (5.8) is a trivial consequence of the extending the argument theorem. However, it is useful to present it here because it has
a very neat expression in terms of distributions. Equation (5.8) simply says that
the marginal distribution of X can be obtained by summing the product of its
conditional distributions given $Y=y$ with the marginal distribution of Y.

From the previous section, we find that an even simpler form can be given:

$$P(X=x|H) = \sum_y P((X=x) \wedge (Y=y)|H).$$ (5.9)

That is, the marginal distribution of X is the sum over y of its joint distribution with Y. It is from this result that the term 'marginal' derives. For, as in Table 5.5 above, we would normally set down a joint distribution of two random variables as a table with rows corresponding to values of one random variable and columns representing values of the other. If we now form marginal totals, i. e. the sums across rows and down columns, we obtain the two marginal distributions. For the employment example, this can be seen in Table 5.6. The bottom row of this table is now the column totals which, from (5.9), is the marginal distribution of X (a copy of Table 5.3). The rightmost column is the row totals, and provides the marginal distribution of Y.

x \ y	0	1	2	3	4	Marginal
0	0.175	0.15	0.09	0.045	0.015	0.475
1	0.05	0.0625	0.06	0.0525	0.045	0.27
2	0.025	0.0375	0.05	0.0525	0.09	0.255
Marginal	0.25	0.25	0.2	0.15	0.15	1.0

Table 5.6. Joint and marginal distributions, employment example

Remember that all these values are the employer's measured probabilities. The joint probabilities in the body of the table were obtained by elaboration from the measurements in Tables 5.3 and 5.4, and the marginal distribution for Y that we have now obtained required further elaboration using (5.9). The result of these two elaborations together is that we have derived an elaborated marginal distribution $P_1(Y=y|H)$ by extending the argument (to include X).

We now consider Bayes' theorem.

Bayes' theorem (distributions)

For any two random variables X and Y, and information H,

$$P(Y=y|(X=x)\wedge H)$$

$$= \frac{P(Y=y|H)P(X=x|(Y=y)\wedge H)}{P(X=x|H)}$$ (5.10)

$$= \frac{P(Y=y|H)P(X=x|(Y=y)\wedge H)}{\sum_{y'} P(Y=y'|H)P(X=x|(Y=y')\wedge H)},$$ (5.11)

where in (5.11) the sum is over all possible values y' of Y.

This follows trivially from Bayes' theorem as presented in Section 3.2. In terms of distributions it says little more than that the conditional distributions of Y given X are obtained by dividing the joint distribution by the marginal distribution of X. However, we know that the principal use of Bayes' theorem is for updating beliefs after acquiring information. Your initial beliefs about the random variable Y are expressed in its *prior distribution* $P(Y=y|H)$. After observing the value of X, Your updated beliefs are in the *posterior distribution* $P(Y=y|(X=x)\wedge H)$. Bayes' theorem shows how the updating is achieved through the *likelihood* of Y based on data $X=x$, the set of probabilities $P(X=x|(Y=y)\wedge H)$ for possible values y of Y.

Notice that the likelihood is *not* a probability distribution. This is an important point. If we hold y fixed and look at $P(X=x|(Y=y)\wedge H)$ for all possible values x of X then we have a distribution, but by holding x fixed and letting y vary we obtain a different set of probabilities. They do not sum to unity (e. g. the rows of Table 5.4 sum to 2.15, 1.4, 1.45, but its columns all sum to 1.0) and indeed their sum is meaningless.

Continuing the employment example, we can derive the conditional distributions from Table 5.6, simply dividing the entries in the table by the corresponding marginal total. Dividing each column by its column total would just produce the original measurements $P_e(Y=y|(X=x)\wedge H)$ in Table 5.4. However, dividing each row by its row total gives Table 5.7, which consists of the conditional probabilities $P_1(X=x|(Y=y)\wedge H)$. Remembering the elaborations which produced Table 5.6, we see that these measurements have been made by elaborating using Bayes' theorem (5.11).

x y	0	1	2	3	4
0	0.37	0.32	0.19	0.09	0.03
1	0.19	0.23	0.22	0.19	0.17
2	0.10	0.15	0.19	0.21	0.35

Table 5.7. Conditional distributions of X given Y, employment example

The figures in Table 5.7 are given only to two decimal places. Each row represents a conditional distribution, and sums to 1.0. The conditional distributions exhibit the kind of behaviour that we would expect in this example. Given that there are no women among the last two applications ($Y=0$), You would not expect many women in the first four, and so on. In fact, Table 5.7 is similar in this respect to Table 5.4, but with the roles of X and Y reversed.

When do extending the argument and Bayes' theorem yield good elaborations? This is a crucial question, which we have tended to ignore in this section. We can answer it partially by saying that the employment example has served to illustrate the mechanics of extending the argument and Bayes' theorem, but it is not an example of good elaboration.

The criteria for good elaboration were spelt out in Chapter 2. When a probability is difficult to measure directly it may be measured more accurately by elaboration if the component probabilities may be measured well directly. This does not apply in the employment example. We used the marginal distribution of X and the conditional distributions of Y given X to elaborate the marginal distribution of Y (extending the argument) and the conditional distributions of X given Y (Bayes' theorem). Both X and Y represent similar things, i.e. the number of women found among a certain number of applicants. It would be no harder to measure $P(Y=y|H)$ directly than $P(X=x|H)$. Nor would it be any harder to measure $P(X=x|(Y=y)\wedge H)$ directly than $P(Y=y|(X=x)\wedge H)$. So in this case the component probabilities are not easier to measure accurately than the target probabilities.

{...........

Elaboration for coherence checking

The above discussion does not mean that the exercise of using extending the argument and Bayes' theorem was entirely useless. Consider Table 5.7 again. Is it really plausible that, if the employer saw two women applicants in the last two applications she would believe that by far the most likely number in the first four applications is four? Certainly she should give a higher probability to $X=4$ given $Y=2$ than if she observed $Y=1$ or $Y=0$, but the value of 0.35 in Table 5.7 is surprisingly high. If she were to measure this probability directly, therefore, she would give a lower value.

The elaborative measurement does not cohere with her direct measurement. Since the elaboration is not particularly accurate, the resolution of the non-coherence is not simply to abandon her direct measurement. There are two approaches which can now be used. The first is to think carefully about all the probabilities she has measured, and by making suitable adjustments bring them into coherence. The second approach is to look for a better elaboration.

...........}

In the next section we present a better example of elaboration by extending the argument and Bayes' theorem. Then in Section 5.5 we return to the employment example to develop an alternative, more accurate elaboration.

5.4 Example: Capture-recapture

How does one determine the number of fish in a lake? The following technique is actually used for this and other problems concerning the sizes of populations of creatures. A net is set up to catch some of the fish. These are marked in some identifiable way and returned to the lake. At a later date another batch of fish are caught. The size of the fish population can then be estimated from seeing how many of the marked fish have been caught in the second sample. If they make up only a very small proportion of the sample then one can reason that marked fish probably make up only a very small proportion of the population generally. This will lead to a large estimate of population size, whereas if most of the second sample is marked then the estimated population size will be smaller.

To keep our example manageable, we shall use quite small numbers. In practice, sample and population sizes would be very much larger. Suppose that six fish have been caught and marked in the first stage, and that ten fish are to be caught in the second stage. The first question is: What is Your probability distribution for X, the number of marked fish which will be found in the second catch?

You could measure Your distribution for X directly, but the natural way to measure these probabilities is to think about how many fish might be in the lake. Let N be the number of fish in the lake, then if $N=15$, 40% of the fish are marked and so You would expect about 40% of the second sample to be marked, i.e. X around four. If $N=30$ You would, by similar reasoning, expect X to be about two. In fact, You can go much further than these intuitive judgements. For if $N=n$ is known Your distribution of X becomes essentially objective. We can then regard the second sample as a deal of 10 from $(6, n-6)$. Thus, if H is Your current information,

$$P(X=x|H \wedge (N=n)) = \binom{6}{x}\binom{n-6}{10-x}/\binom{n}{10} . \tag{5.12}$$

For instance, Table 5.8 shows Your distribution for X given $H \wedge (N=20)$.

x	0	1	2	3	4	5	6
Probability	0.005	0.065	0.244	0.372	0.244	0.065	0.005

Table 5.8. Objective conditional distribution $P(X=x|H \wedge (N=20))$ of second marked fish catch given population size 20

Of course, You do not know N, and the distribution You require is the (marginal) distribution of X given H. This can be obtained by extending the argument to include N, allowing You to use the highly accurate component probabilities (5.12). The other components comprise Your distribution of N given H.

Suppose now that You are confident that N is between 15 and 30. This restricts the number of possible values for N, but there remain 16 probabilities for You to measure in determining Your distribution for N. It is not a simple task to measure these probabilities with reasonable accuracy. In fact, we shall return to this matter in Section 5.5. For our present purposes, we shall just suppose that the probabilities have been measured as given in Table 5.9.

n	15	16	17	18	19	20	21	22
$P_1(N=n\|H)$	.01	.02	.04	.07	.11	.13	.13	.12
n	23	24	25	26	27	28	29	30
$P_1(N=n\|H)$	.11	.09	.06	.04	.03	.02	.01	.01

Table 5.9. Measured marginal probabilities of fish population size

Multiplying these measurements by the objective probabilities (5.12) gives Your (measured) joint distribution for N and X given Your information H. It could be written as a table of sixteen rows and seven columns. Summing over the possible values of N (the rows of the table) gives Your marginal distribution for X, shown in Table 5.10. The computations which have been performed in deriving these values consist of extending the argument to include the random variable N.

x	0	1	2	3	4	5	6
$P_1(X=x\|H)$	.014	.101	.272	.341	.209	.058	.005

Table 5.10. Measured marginal distribution of second marked fish catch

This is quite a good use of extending the argument. Although the marginal distribution of X comprises only seven probabilities, they are difficult to measure directly. The reason for the difficulty is that Your uncertainty about X is naturally made up of two quite different sources of uncertainty. One is uncertainty about the population size and the second is induced by the vagaries of the second catch. Elaborating by extending the argument separates these two components. Furthermore, the second source of uncertainty (the catch mechanism) has objective probabilities.

It is now a simple matter to apply Bayes' theorem to solve a far more important problem: given a particular observed number of marked fish in the second catch, what can You infer about N? This is a natural and powerful application of Bayes' theorem. Your prior distribution is a subjective, measured distribution, presented in Table 5.9. But the likelihood, equation (5.12), is objective. Table 5.11 presents all the possible posterior distributions of N. Each column gives Your posterior distribution $P_1(N=n|(X=x)\wedge H)$ for a given number x of marked fish caught.

These posterior distributions behave very much as we described in the first paragraph of this section. The larger x is, the smaller You will then believe the population to be. In particular, the most probable value of N for each x (the largest value in each column) is shown in italics. We can see that if You catch no marked fish in the second catch You will believe that the most likely value of N is 24, whereas if You catch all six of the marked fish the most probable value of N is 19. There is a steady transition between these two extremes.

x n	0	1	2	3	4	5	6
15	0.000	0.000	0.002	0.007	0.020	0.044	0.079
16	0.000	0.002	0.006	0.018	0.038	0.066	0.098
17	0.002	0.007	0.019	0.040	0.068	0.099	0.127
18	0.008	0.021	0.044	0.074	0.106	0.132	0.149
19	0.024	0.050	0.085	0.120	0.146	*0.160*	*0.160*
20	0.050	0.083	0.117	*0.141*	*0.151*	0.147	0.132
21	0.078	0.109	0.131	0.139	0.132	0.115	0.095
22	0.105	0.126	*0.132*	0.124	0.107	0.085	0.063
23	0.132	*0.138*	0.129	0.110	0.085	0.062	0.043
24	*0.141*	0.132	0.111	0.086	0.061	0.041	0.026
25	0.119	0.100	0.077	0.054	0.034	0.022	0.013
26	0.098	0.075	0.052	0.034	0.021	0.012	0.007
27	0.088	0.062	0.040	0.024	0.014	0.008	0.004
28	0.069	0.045	0.027	0.015	0.008	0.004	0.002
29	0.040	0.024	0.014	0.007	0.004	0.002	0.001
30	0.046	0.026	0.014	0.007	0.003	0.001	0.001

Table 5.11. Measured posterior distributions of fish population size

The effect of observing X is even more marked when we compare Your posterior probability that N is less than 20 with Your prior probability, which is $P_1(N < 20 | H) = 0.01 + 0.02 + 0.04 + 0.07 + 0.11 = .25$. Table 5.12 shows how Your posterior probabilities can be anything from $P_1(N < 20 | (X=0) \wedge H) = .034$ to $P_1(N < 20 | (X=6) \wedge H) = .613$.

x	0	1	2	3	4	5	6
Probability	0.034	0.080	0.156	0.259	0.378	0.501	0.613

Table 5.12. Measured posterior probabilities $P_1(N < 20 | (X=x) \wedge H)$
for fewer than 20 fish in lake

5.5 Example: job applications

The employment example of Section 5.3 considered two random variables, X = number of women applicants in the first four out of six job applications, and Y = number of women in the remaining two applications. Because X and Y refer to similar things, it is no more difficult or easy to measure directly the conditional probabilities of X given Y than of Y given X. Therefore Bayes' theorem is not a good elaboration of either. Contrast this with the fish-catch example. It is much easier to measure the conditional probabilities of X given N than of N given X. Indeed, the former probabilities can be measured objectively.

The division between the first four and the last two applications is arbitrary. For the employer it is just as easy to consider the random variable $Z=X+Y$, the total number of women applicants. Not only is this a natural thing to do, but we shall find that it enables us to make objective probability statements in this example too.

We therefore suppose that the employer measures her distribution for Z (given her information H before seeing any applications) as in Table 5.13.

z	0	1	2	3	4	5	6
$P_2(Z=z\,|\,H)$	0.10	0.20	0.20	0.20	0.15	0.10	0.05

Table 5.13. Employers probabilities for total women applicants

Now consider the original two random variables X and Y. In either case, conditioning on the value of Z yields objective probabilities. For, given $Z=z$, X is equivalent to the number of successes in a deal of four from $(z, 6-z)$ and Y is equivalent to the number of successes in a deal of two from $(z, 6-z)$. The conditional distributions of Y given Z are presented in Table 5.14.

z \ y	0	1	2	3	4	5	6
0	1	$2/3$	$2/5$	$1/5$	$1/15$	0	0
1	0	$1/3$	$8/15$	$3/5$	$8/15$	$1/3$	0
2	0	0	$1/15$	$1/5$	$2/5$	$2/3$	1

Table 5.14. Objective probabilities $P(Y=y\,|\,(Z=z)\wedge H)$

Using Tables 5.13 and 5.14 we can derive the (measured) joint distribution of X and Y by the following elaboration.

$$P((X=x)\wedge(Y=y)\,|\,H) = P((Z=x+y)\wedge(Y=y)\,|\,H)$$

$$= P(Z=x+y\,|\,H)\,P(Y=y\,|\,(Z=x+y)\wedge H)$$

Applying this for all possible values of X and Y yields Table 5.15.

x \ y	0	1	2	3	4	$P_2(Y=y\,|\,H)$
0	0.10	0.13	0.08	0.04	0.01	0.36
1	0.07	0.11	0.12	0.08	0.03	0.41
2	0.01	0.04	0.06	0.07	0.05	0.23
$P_2(X=x\,|\,H)$	0.18	0.28	0.26	0.19	0.09	1.00

Table 5.15. Measured joint and marginal distributions,
the employment example

The corresponding table in Section 5.3, Table 5.6, was based on the twenty probability measurements presented in Tables 5.3 and 5.4. Table 5.25 relies only on the seven measured probabilities in Table 5.13, and is otherwise objective. The new elaboration should be substantially more accurate.

It is now a simple matter to obtain conditional distributions. For example, the conditional distributions of X given Y are shown in Table 5.16.

x	0	1	2	3	4
y					
0	0.28	0.36	0.22	0.11	0.03
1	0.17	0.27	0.29	0.20	0.07
2	0.04	0.17	0.26	0.31	0.22

Table 5.16. Conditional distributions $P_2(X=x|(Y=y)\wedge H)$, the employment example

The kind of elaboration used in this section is actually very important, and is dealt with in detail in Chapter 8. We postpone further discussion of these elaboration techniques until then.

Exercises 5(b)

1. Steel cables are supplied to a contractor. They are claimed to be made from a high-quality steel, but the contractor suspects that they might be of a poorer quality. He decides to examine ten cables. Let X be the number of faulty cables found, let $E \equiv$ 'high quality steel used', and let H be his current information. The table below shows the contractor's measured distributions for X given $H \wedge E$ and given $H \wedge (\neg E)$. In addition, he measures $P_2(E|H)=0.7$. Elaborate his distribution for X given H.

x	0	1	2	3	4
$P_1(X=x|H\wedge E)$	0.6	0.3	0.1	0.0	0.0
$P_1(X=x|H\wedge(\neg E))$	0.3	0.3	0.2	0.1	0.1

2. An automatic sensor counts damaged cells on microscope slides. With probability 0.5 it gives the correct count, but it may also give a count which is one more or one less than the true count. There is probability 0.25 for each of these kinds of error. Let X be the true count and Y the sensor's count. Based on experience with such counts, a biologist gives the following distribution for X.

x	1	2	3	4	5	6	7	8
$P_1(X=x|H)$	.06	.17	.19	.18	.16	.14	.08	.02

Derive the biologist's joint distribution of X and Y, and her marginal distribution of Y. Given an observed sensor count of $Y=2$ cells, what should the biologist now believe about X?

3. Suppose that in the capture-recapture example of Section 5.4, You find $X=1$ marked fish in the second catch. You mark the nine other caught fish and return them all to the lake. You now make a third catch, of two fish. Let Y be the number of marked fish in the third catch and let $H_1 \equiv H \wedge (X=1)$ be Your current information. Calculate the (objective) distribution of Y given $H_1 \wedge (N=20)$. Derive Your distribution of Y given H_1. (N.B. You should take Your distribution of N given H_1 from Table 5.11.)

4. Following Exercise 3, derive Your distribution of N given $H_1 \wedge (Y=0)$.

5.6 Mean and standard deviation

Suppose that, in the capture-recapture example of Section 5.4, You observe $X=2$ marked fish in the second catch. Then Your posterior distribution for N is given as the third column in Table 5.11. As an expression of Your beliefs about N this comprises quite a full and detailed statement, but often one requires simpler kinds of statements, either instead of or as a supplement to a full probability distribution. In particular, it is frequently valuable to be able to state a *typical* value of a random variable. We tried to do this in Table 5.11 by italicizing the largest value in each column. Thus given $X=2$ the most probable value of N is 22, with $P_1(N=22 | (X=2) \wedge H) = 0.132$. The most probable value in a distribution is called its *mode*, but as a 'typical' value the mode is not always very helpful. One of its drawbacks is noticeable in this example because $P_1(N=21 | (X=2) \wedge H) = 0.131$ is very close to the modal probability of 0.132. The discrepancy is certainly within the bounds of measurement error, so it is not clear which value, $N=21$ or $N=22$, should be regarded as the mode.

There are many ways of defining a 'typical' value for a random variable, one of which is the mode. In general, they are known as *measures of location.*

Measures of location

Any quantity which in some sense represents a 'typical' value of a random variable may be called a measure of location. One measure of location is the mode, which is the value of the random variable having the highest probability. A measure of location is useful as a *summary*, or simplified description of a distribution.

Each such measure has its own advantages and disadvantages. Each expresses the notion of 'typical' in a different way. To discuss even a few of them in any detail here would be too much of a distraction from the main course of our argument. Instead we shall present, with a minimum of comment, one of the more useful candidates.

The *mean*, or *expectation*, of a random variable X, given H, is computed as follows. Each possible value x of X is multiplied by the corresponding probability $P(X=x|H)$ and the resulting products are all added together. Thus the expectation of Y given H in the last section, using the marginal probabilities in Table 5.15, is

$$(0 \times 0.36) + (1 \times 0.41) + (2 \times 0.23) = 0.87 .$$

The modal value is 1, but the expectation is lower, reflecting the fact that $Y=0$ is more likely than $Y=2$. One obvious drawback of a random variable's expectation, thinking of it as a typical value, is that it is generally not a possible value.

Mean, or expectation

The expectation, or mean, of a random variable X given H is denoted by $E(X|H)$ and defined to be

$$E(X|H) = \sum_x x\, P(X=x|H) , \qquad (5.13)$$

where the sum is over all possible values x of X. The mean is a measure of location.

The expectation $E(X|H)$ is typical in the sense of being a kind of average value of X. It is a *weighted average*, with each possible value being weighted according to its probability given H.

{...........

Weighted average

In everyday use, the 'average' of a set of numbers $x_1, x_2, \ldots, x_n$ is

$$\bar{x} = n^{-1} \sum_{i=1}^{n} x_i = \sum_{i=1}^{n} (n^{-1}) x_i .$$

This is a special case of what is called a weighted average, which has the form

$$x^* = \sum_{i=1}^{n} a_i x_i ,$$

where the *weights* a_i are any non-negative constants which add to one, i.e.

$$\sum_{i=1}^{n} a_i = 1 .$$

In the case of the ordinary average the a_is are all n^{-1} which, since there are n of them, add to one as required. The ordinary average is a weighted average in which all the x_is receive the same weight. Notice that whatever weights we use, if every x_i is greater than or equal to some constant b, then

$$x^* \geq \sum_{i=1}^{n} a_i b = b \sum_{i=1}^{n} a_i = b .$$

Similarly, if every x_i is less than or equal to some constant c then $x^* \leq c$. In particular, any weighted average of a set of numbers lies between the smallest and the largest of those numbers. It is also obvious that if one x_i receives a weight close to one then x^* will be close to that x_i, whereas if the weights are all very similar then x^* will be close to $\bar{x}$.

..........}

The usual distinction between true and measured values applies to expectations. We have defined the true value $E(X|H)$ in (5.13) in terms of the true probabilities $P(X=x|H)$, but in practice we can only have a measured value $E_m(X|H)$. One obvious way of measuring $E(X|H)$ is to substitute measured probabilities $P_m(X=x|H)$ in (5.13). This is the natural elaboration, but we shall also consider a more direct measurement in Section 5.7.

{..........

Examples

From Table 5.11 we can compute the expected value of N given $H \wedge (X=x)$ for each x. For instance, $E_1(N|(X=6) \wedge H)$ is

$$(15 \times 0.079) + (16 \times 0.098) + ... + (30 \times 0.001) = 18.998 .$$

The table shows how $E_1(N|(X=x) \wedge H)$ falls smoothly with x.

x	0	1	2	3	4	5	6
$E_1(N|(X=x) \wedge H)$	24.4	23.4	22.4	21.5	20.6	19.7	19.0

The prior mean is

$$E_1(N|H) = (15 \times 0.01) + (16 \times 0.02) + ... + (30 \times 0.01) = 21.66$$

(from Table 5.9).

..........}

In the idea of an expectation we begin to see the value of using random variables. A partition is just an unstructured set of propositions. But if each is associated with a number, a possible value of a random variable, then structure is created and we can do new things with the numbers themselves.

After a measure of location the next most useful summary statement we can make about a random variable and its distribution is to provide a measure of *dispersion*. A dispersion measure describes, roughly speaking, how far from its 'typical' value the random variable might 'typically' be. For instance, returning to the distribution of N given $(X=2) \wedge H$ in Section 5.6, whilst N is most likely to be 21 or 22, it could almost as easily be 19, 20, 23, 24 or 25, but the probability of it being further from 21 or 22 than this is rather low.

Measure of dispersion

A measure of dispersion for a distribution is an indication of how far a random variable might probably be from a corresponding measure of location.

Just as there are many measures of location, there are many measures of dispersion. We shall again present just one such measure – the *standard deviation*.

The standard deviation of a random variable X given information H is computed as follows. First we calculate the expectation $E(X|H)$. Then for each possible value x of X we subtract $E(X|H)$ from x and square the result, obtaining

$$d(x) \equiv \{ x - E(X|H) \}^2 .$$

Next we multiply each $d(x)$ by $P(X=x|H)$ and sum over all possible x giving

$$var(X|H) \equiv \sum_x d(x) P(X=x|H) .$$

Finally, the standard deviation of X given H is the square root of this quantity. That is, the standard deviation of X is the square root of the average squared distance of X from its mean $E(X|H)$. The square of the standard deviation, $var(X|H)$, is called the *variance* of X given H.

Variance and standard deviation

The variance of a random variable X given H is denoted by $var(X|H)$ and is defined to be

$$var(X|H) \equiv \sum_x \{ x - E(X|H) \}^2 P(X=x|H) , \tag{5.14}$$

where the sum is over all possible values x of X. It represents the expected squared deviation of X from its mean.

The standard deviation of X given H is denoted by $sd(X|H)$, and defined by

$$sd(X|H) \equiv \{ var(X|H) \}^{\frac{1}{2}} . \tag{5.15}$$

The standard deviation is a measure of dispersion.

The variance is a weighted average of the squared deviations $d(x)$. Notice that since each $d(x) \geq 0$ we must have $var(X|H) \geq 0$, and therefore $sd(X|H) \geq 0$. Dispersion measures are generally non-negative quantities. Zero dispersion naturally means that a random variable is necessarily exactly equal to its location measure. This is true of the standard deviation, for we can only have

$sd(X|H)=0$ or $var(X|H)=0$ if every $d(x)$ for which $P(X=x|H)\neq0$ is zero, which implies that the only possible value of X given H is $E(X|H)$. We call a random variable with zero dispersion *degenerate*, because it is not genuinely random. It effectively has only one possible value. With the exception of degenerate random variables, $sd(X|H)$ is always strictly positive.

{...........
Example
Working again from Table 5.11, let us compute the standard deviation of N given $H\wedge(X=3)$. We found $E_1(N|(X=3)\wedge H)=21.5$ in the previous example. The possible values of N are 15, 16, 17 ,..., 30, and the corresponding values of $d(n)$ are $d(15)=(15-21.5)^2=(-6.5)^2=42.25$ and so on. Therefore

$$var_1(N|(X=3)\wedge H) = (0.007\times42.25)+...+(0.007\times72.25)$$

$$= 8.082 .$$

$$\therefore sd_1(N|(X=3)\wedge H) = 8.082^{1/2} = 2.84 .$$

Notice that the same distinction between true and measured quantities applies to variances and standard deviations. Table 5.11 gives measured probabilities so these are also measured quantities.

Applying the same calculations to the other columns of Table 5.11 gives the standard deviations of the various posterior distributions of numbers of fish as follows

x	0	1	2	3	4	5	6
$sd_1(N|(X=x)\wedge H)$	2.82	2.88	2.89	2.84	2.78	2.69	2.58

...........}

Your standard deviation of X is a measure of Your degree of uncertainty about X. If X has a very small standard deviation then You believe that X will almost certainly lie close to its mean. Therefore it is highly probable to take one of a small number of different values. In this case, Your information about X is strong, i.e. Your uncertainty is low. If, on the other hand, X's standard deviation is high then You believe it could be very far from its expectation (either higher or lower) and therefore You are very uncertain about its likely value.

{...........
Example
Continuing the above example, Your posterior uncertainty about N, the number of fish in the lake, varies little with x. If, however, we compute the standard deviation of Your prior distribution (Table 5.9) we find $sd_1(N|H)=3.02$. Observing the number of marked fish in the second catch reduces Your uncertainty. Observing $X=6$ causes the greatest reduction of

Your uncertainty, but even in the worst instance, $X=2$, Your standard deviation of N is reduced from 3.02 to 2.89. This is a small reduction because the second catch is only small. In a larger second catch the observation of X would be more informative.

..........}

In general, extra information reduces uncertainty. We shall prove a theorem to this effect in Chapter 6, Section 6.7.

Exercises 5(c)

1. For the distributions of X and Y given in Exercise 5(a)3, find $E(X|H)$ and $E(Y|H)$. Find also $E(Z|H)$, where $Z=X+Y$, using the distribution of Z obtained in Exercise 5(a)3. Confirm that $E(Z|H)=E(X|H)+E(Y|H)$. [This is an instance of a general result proved in Chapter 6.]

2. Compute the variances and standard deviations of the distributions of X, Y and Z in Exercise 5(a)3.

3. Find means and standard deviations for the distributions of X given H, and X given $H \wedge (Y=2)$ from Exercise 5(b)2. (Confirm that learning $Y=2$ reduces Your standard deviation of X.)

4. Calculate expectations and standard deviations for the random variables T, L, S and P in Exercise 5(a)1.

5.7 Measuring distributions

The distribution of a random variable X, given H, consists of the probabilities $P(X=x|H)$ for all possible values x. If You measure each of these probabilities separately, then the result can be called a *measured distribution*. If the number of possible values is at all large then it is a substantial task to measure each probability separately. One problem is that with many possible values the probability of X taking any given value is likely to be small. A measurement error of plus or minus 0.01 is unrealistically small for direct measurement, yet it is obviously dangerously large if the probabilities to be measured are also around 0.01. Furthermore, the errors accumulate. The sum of probabilities in a distribution is always unity, but adding the measured probabilities in a distribution with many possible values can yield a total far from one, representing a large cumulative error. Removing this non-coherence is difficult because there are many probabilities which could be adjusted and typically none of them can be regarded as being much more accurate than any others. Fortunately, considerable assistance in measuring distributions is provided by the notion of *shape*.

To see what we mean by the shape of a distribution consider the prior distribution of N in the capture-recapture example, Table 5.9. Notice that the measured probabilities fall away steadily on either side of $N=20$ and 21. At the extremes, the probability falls to 0.01 at $N=15$, 29 and 30, until it dies to zero effectively outside this range. This is a very common shape for a probability distribution. The most probable values are in the 'centre', and values further from the centre are increasingly improbable. A distribution like this is called *unimodal*, which is the most common shape.

Shape

The shape of a distribution is a description of the way in which $P(X=x|H)$ changes as the possible value x increases. In addition to measures of location and dispersion, descriptions of shape are valuable summaries of a distribution.

If $P(X=x|H)$ increases with x until the mode is reached and then decreases with x, the distribution of X given H is called unimodal.

Looking more closely at the distribution of N in Table 5.9 we see that the probabilities decrease on the right of the mode slightly more slowly than they increase on the left of the mode. We would describe this aspect of its shape as 'slight skewness to the right'.

Skewness

A unimodal distribution may be further summarized by a measure of skewness, which is the amount of discrepancy between the rate at which $P(X=x|H)$ increases before the mode, and the rate at which it decreases thereafter. If it increases faster than it decreases it is called skewed to the right, whereas if it decreases faster than it increases it is called skewed to the left. The greater the difference between these rates the higher the degree of skewness.

A distribution which is symmetric about its mode has no skewness.

Whilst it is important to be aware of skewness in general, we do not consider any measures of skewness here.

Shape was used in deriving Table 5.9 in the following way. First, $N=20$ and 21 were decided upon as the most likely values. Then it was decided that values lower than $N=17$ or higher than $N=27$ were highly improbable. Next, the shape was judged to be unimodal. Finally some probabilities were selected to fit these requirements (and to sum to one). The resulting probabilities, and the distribution as a whole, have clearly been measured, but in a way very different from direct measurement or any elaboration that we have met previously. Yet it is still a kind of elaboration. The components of the elaboration are not themselves probabilities. Instead they are, in a crude sense, measurements of location, dispersion and shape. Choosing $N=20$ and 21 as most likely values to form the 'centre' of the distribution corresponds to choosing a location around $N=20$ or 21. Saying that values outside the range 17 to 27 will be highly improbable corresponds to a measurement of dispersion (and an indication of slight skewness to the right). Finally, the unimodal shape was an explicit 'measurement'.

This way of measuring a distribution is typically more accurate and more efficient than direct measurement of each separate probability. We can see how higher accuracy has been achieved in this case, because it is hard to vary any of the probabilities up or down by more than about 0.02 without disturbing the shape or violating the judgements of location and dispersion. The greater efficiency applies in the sense that we have made only three 'measurements' – location, dispersion, shape – instead of 16 direct probability measurements. The gain is even more dramatic if we think of a more realistic version of the same problem. A real lake might contain hundreds or thousands of fish, and so there would be hundreds or thousands of possible values for N. Yet these same three measurements – location, dispersion, shape – would still enable a probability distribution to be determined in which individual probabilities are measured to high accuracy.

Elaborated distributions

Probability distributions can be measured using a kind of elaboration in which the components are various summaries of the distribution. The most frequently used summaries are descriptions of shape and measures of location and dispersion. The elaboration consists of letting $P(X=x|H)$ be any suitable values which agree with the stated summaries, and which sum to one. This form of elaboration can be highly accurate and efficient if there are many possible values of the random variable.

This whole process can be exploited even more fully with the aid of another useful idea. Probabilists, mathematicians and statisticians have devised a great many *standard distributions* which will automatically give sensible probability values agreeing with the stated summary measurements.

For example, suppose that a physicist wishes to express his current beliefs about the half-life of a certain radioactive material. Let H be his knowledge and let X be the half-life of the material, to the nearest year. He measures his distribution for X given H as follows. He estimates that the half-life will be around 52 years, representing a 'typical' value. His distribution should be unimodal and, although values between 45 and 60 years are not unreasonable, he regards values as low as 40 or as high as 65 as distinctly unlikely. We shall see in the next section that there is a standard distribution, known as the Poisson distribution with mean 52, which fits these statements of beliefs. The physicist therefore simply measures his distribution for X given H as being the Poisson distribution with mean 52. Specifically (see Section 5.8), his measured probabilities are now given by the formula

$$P_1(X=x|H) = 52^x \exp(-52)/x! . \tag{5.16}$$

Computing (5.16) for a few different x values, we obtain Table 5.17 below, and the whole distribution is shown graphically in Figure 5.1.

x	43	45	53	52	55	60	65
$P_1(X=x|H)$	.014	.036	.054	.055	.049	.029	.011

Table 5.17. Selected probabilities from a Poisson distribution
with mean 52

Notice from Figure 5.1 that the distribution has a very smoothly unimodal shape. A considerable number of probabilities have been measured quickly by using the device of standard distributions.

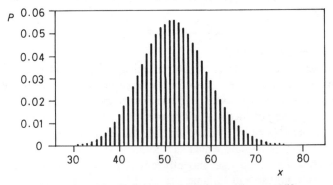

Figure 5.1. The Poisson distribution with mean 52

{...........

The exponential function
The expression $\exp(-52)$ in equation (5.16) denotes the exponential function. In general, $\exp(x)$ is e^x, i.e. the mathematical constant $e=2.71828...$ raised to the power x. We use the $\exp(\ )$ notation particularly when the power itself is a complicated expression. For instance $\exp(-x^2-1)$ is easier to read than e^{-x^2-1}

...........}

Using standard distributions

The elaborative measurement of distributions, by choosing convenient probabilities which agree with measurements of location, dispersion, shape or other summaries, is further simplified by the existence of a variety of standard distributions. It is often possible simply to select a standard distribution having the desired summaries, whereupon convenient probabilities are automatically generated from a formula.

The cumbersome phrase 'X given H has the Poisson distribution with mean 52' is denoted simply by

$$X|H \sim \text{Po}(52) . \qquad (5.17)$$

There are two new items of notation in (5.17). First, the symbol '$\sim$' means 'has the same distribution as'. In general we write $X|H \sim Y|H*$ to denote that, on information H, X has the same distribution as that of Y given information $H*$.

The equi-distribution symbol

The symbol $\sim$ is read 'has the same distribution as'. Thus

$$X|H \sim Y|H*$$

means that X and Y have the same possible values, and that for any such value z,

$$P(X=z|H) = P(Y=z|H*) .$$

{...........

Example: independence

The definition of independence in Section 5.2 says that X and Y are independent given H if $P(X=x|H) = P(X=x|(Y=y)\wedge H)$ for all x and y. We could now express this as

$$X|H \sim X|(Y=y)\wedge H$$

for all possible values y of Y. We could go on to say that a random variable X and a proposition E are independent given H if

$$X|H \sim X|H \wedge E . \qquad (5.18)$$

In other words, knowledge of E does not change Your beliefs about X.

...........}

With this interpretation of '~', we can see that the symbol Po(52) in (5.17) represents a random variable. It is a random variable which is *defined* to have the Poisson distribution with mean 52. Information cannot alter a definition, so Po(52) has this distribution, by definition, for all H. Because information is irrelevant to the distribution of Po(52) it is suppressed in (5.17). All standard distributions are identified by a standard random variable, having a short symbolic name like Po(52), which has that distribution by definition. The random variable Po(1), for instance, is defined to have the Poisson distribution with mean one. In general, Po(m) is a random variable defined to have the Poisson distribution with mean m. There is a complete family of Poisson distributions, the *parameter m* indexing the members of the family, and being identified with the mean of the distribution. Standard distributions are readily grouped into families like this, each family indexed by one or more parameters.

As a final comment on the notation introduced by (5.17), consider the distinction between true and measured distributions. The equi-distribution symbol ~ can have a subscript to denote a measured distribution. Since ~ relates two distributions, one each side of it, subscripts may be attached to ~ on either side, or on both. Thus, in the half-life example, the physicist's distribution for X given H is measured, and instead of (5.17) we should use

$$X|H \underset{1}{\sim} Po(52) , \qquad\qquad (5.19)$$

the subscript 1 before the ~ showing that we are referring to a measured distribution for $X|H$. Clearly, the distributions of standard random variables like Po(52) are true distributions, so there is no subscript to the right of ~ in (5.19).

5.8 Some standard distributions

In this section we will catalogue the most important families of standard distributions. In each case we will

(a) define the distributions and associated standard random variables,
(b) give their means and variances (often without proof),
(c) describe their shapes, and
(d) illustrate shape by diagrams of selected distributions in the family.

Each family will begin with a displayed definition dealing with parts (a) and (b). Then in normal text we will consider shape and any new matters which arise.

The uniform family

The random variable $U(n)$ is defined to have the uniform distribution on the first n positive integers, whose probabilities are

$$P(U(n)=x) = n^{-1},$$

and whose possible values are $x = 1, 2, \ldots, n$.

$$E(U(n)) = (n+1)/2.$$

$$var(U(n)) = (n^2 - 1)/12.$$

Notice that probabilities, expectation and variance omit the usual '$|H$' because the definition applies for all H. Remember that we called two propositions logically mutually exclusive if they were mutually exclusive for all H, and have made similar use of the word 'logical' in other contexts. We can regard a standard random variable's distribution as a logical consequence of its definition.

{...........

Proof

The mean of $U(n)$ is easily obtained from the well-known result

$$\sum_{x=1}^{n} x = n(n+1)/2,$$

$$\therefore \sum_{x=1}^{n} x P(U(n)=x) = \frac{1}{n} \sum_{x=1}^{n} x = (n+1)/2.$$

Its variance may be obtained from the result

$$\sum_{x=1}^{n} x^2 = n(n+1)(2n+1)/6.$$

$$\therefore var(U(n)) = \sum_{x=1}^{n} (x-(n+1)/2)^2 P(U(n)=x)$$

$$= \frac{1}{n} \sum_{x=1}^{n} \{x^2 - (n+1)x + (n+1)^2/4\}$$

$$= \frac{1}{n} \sum_{x=1}^{n} x^2 - \frac{n+1}{n} \sum_{x=1}^{n} x + \frac{(n+1)^2}{4}$$

$$= \frac{(n+1)(2n+1)}{6} - \frac{(n+1)^2}{2} + \frac{(n+1)^2}{4}$$

$$= \frac{n+1}{12} \{2(2n+1) - 3(n+1)\} = (n+1)(n-1)/12.$$

However, as with all variances presented in this section, it may be obtained rather more easily using the theory of expectations given in Section 6.3.

...........}

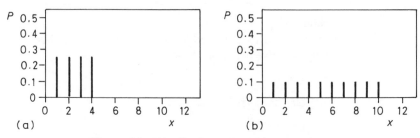

Figure 5.2. Distributions of (a) U(4), (b) U(10)

Turning to shape, the uniform distributions are 'shapeless'. Their probabilities neither rise nor fall. Figure 5.2 shows two typical examples.

The uniform distributions arise from random variables defined on equi-probable partitions. If You have no information about a random variable except that it has possible values $1, 2, \ldots, n$, then its distribution will be that of $U(n)$.

The binomial family

The random variable $\text{Bi}(n,p)$ is defined to have the binomial distribution with parameters n and p, whose probabilities are

$$P(\text{Bi}(n,p)=x) = \binom{n}{x} p^x (1-p)^{n-x} ,$$

and where the possible values are $x=0, 1, 2, \ldots, n$. The parameter p lies in the range $[0, 1]$.

$$E(\text{Bi}(n,p)) = np .$$

$$var(\text{Bi}(N,p)) = np(1-p) .$$

{...........

Proof

We give only the derivation of the mean. By definition,

$$E(\text{Bi}(n,p)) = \sum_{x=0}^{n} x \binom{n}{x} p^x (1-p)^{n-x} .$$

Notice first that the term in the sum for $x=0$ is itself zero, so that $\sum_{x=0}^{n}$ can be altered to $\sum_{x=1}^{n}$. Next consider

$$x \binom{n}{x} = \frac{x\, n!}{x!\,(n-x)!} = \frac{n!}{(x-1)!\,(n-x)!} = \frac{n\,(n-1)!}{(x-1)!\,(n-x)!}$$

$$= n \binom{n-1}{x-1} .$$

$$\therefore E(\text{Bi}(n,p)) = \sum_{x=1}^{n} n \binom{n-1}{x-1} p^x (1-p)^{n-x} \tag{5.20}$$

$$= np \sum_{x=1}^{n} \binom{n-1}{x-1} p^{x-1}(1-p)^{n-x} .$$

Now the sum on the right in (5.20) is the sum of all the probabilities in the distribution of $\text{Bi}(n-1,p)$, and therefore equals one. This may be seen more clearly if we write $x^* = x-1$ and $n^* = n-1$. Then the range $x=1$ to n becomes $x^* = 0$ to n^* and we have

$$\sum_{x=1}^{n} \binom{n-1}{x-1} p^{x-1}(1-p)^{n-x} = \sum_{x^*=0}^{n^*} \binom{n^*}{x^*} p^{x^*}(1-p)^{n^*-x^*} .$$

The sum of the probabilities in a distribution is always one. Often, when proving some result about a distribution, we have a summation to perform which becomes easy when it is manipulated (like (5.20)) into the sum of probabilities in some other distribution.

..........}

We have already seen that if X is the number of successes in n binary trials with probability p of success in each trial, given H, then $X|H \sim \text{Bi}(n,p)$.

For moderate values of p, the binomial distributions are unimodal. If $p = \frac{1}{2}$ the distribution is symmetric about the mean $n/2$. For $p < \frac{1}{2}$ it is skewed to the right and for $p > \frac{1}{2}$ it is skewed to the left. The skewness increases as p moves further from $\frac{1}{2}$. If p becomes sufficiently close to 0 or 1 the skewness will be such that the mode is at $x=0$ or $x=n$ respectively. Then the distribution is not strictly unimodal because either the increasing or the decreasing section has disappeared. Such a distribution is said to be J-shaped. These shapes are all shown in Figure 5.3.

The hypergeometric family
The random variable $\text{Hy}(N,R,n)$ is defined to have the hypergeometric distribution with parameters N, R and n, whose probabilities are

$$P(\text{Hy}(N,R,n)=x) = \binom{R}{x}\binom{N-R}{n-x} / \binom{N}{n} .$$

The possible values of x range from $\max(0, n+R-N)$ to $\min(R, n)$. The parameters satisfy $N \geq R$ and $N \geq n$.

$$E(\text{Hy}(N,R,n)) = nR/N .$$

$$var(\text{Hy}(N,R,n)) = n(N-n)(N-R)R/\{N^2(N-1)\} .$$

We have met hypergeometric distributions before. If, in a deal of n from $(R, N-R)$, given H, X is the number of propositions of type 1 that are true, then $X|H \sim \text{Hy}(N,R,n)$. We also know, from Section 4.11, that as $N \to \infty$ and $R \to \infty$ so that R/N is fixed at $R/N=p$, the distribution of $\text{Hy}(N,R,n)$ tends to

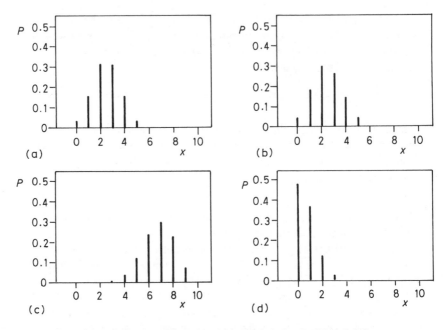

Figure 5.3. Distributions of (a) Bi(5, 0.5), (b) Bi(7, 0.35),
(c) Bi(9, 0.75), (d) Bi(7, 0.1)

that of Bi(n, p). For large N, therefore, the hypergeometric distributions will
have very similar shapes to the binomial distributions. For smaller N, they still
have the same general shapes, i.e. unimodal or J-shaped, with symmetry when
$N = 2R$ and varying degrees of skewness otherwise. However, the difference is
that they have lower dispersion than the corresponding binomial distributions,
and as N decreases to n the variance decreases to zero. This behaviour is
shown in Figure 5.4. Figure 5.4(a) is very like the corresponding binomial distribution in Figure 5.3(c), but the other diagrams in Figure 5.4 show steadily less
dispersion as N decreases. We can think of the hypergeometric family as an
enhancement of the binomial family through the addition of a third parameter.

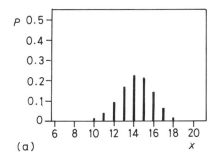

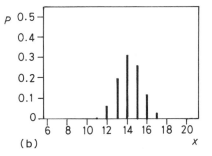

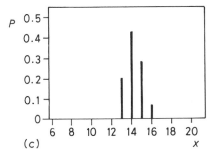

Figure 5.4. Distributions of (a) Hy(100, 75, 19), (b) Hy(32, 24, 19), (c) Hy(24, 18, 19)

The geometric family

The random variable Ge(p) is defined to have the geometric distribution with parameter p, whose probabilities are

$$P(\text{Ge}(p)=x) = p(1-p)^x ,$$

and where the possible values are $x = 0, 1, 2, \ldots$. The parameter p lies in [0,1].

$$E(\text{Ge}(p)) = (1-p)/p .$$

$$\text{var}(\text{Ge}(p)) = (1-p)/p^2 .$$

The possible values of Ge(p) are given as $0, 1, 2, \ldots$. There is no upper bound. Unlike other random variables that we have considered until now, Ge(p) does not have a finite number of possible values. The implications of this are both theoretical and practical.

On a practical level, it is not unrealistic to consider an unbounded random variable. Let X be the number of years until the next eruption of a volcano in South America. There is a long geological fault running down South America, and it is inconceivable that there will never be another volcanic eruption there. It is highly improbable, but not impossible, that we will wait ten years before the next eruption. It is even more improbable, but still not impossible, that we might wait 100 years. This is a random variable to which we cannot logically assign an upper bound.

Unbounded random variables introduce the minor technical complication of infinite sums. For instance, to verify that the sum of the probabilities in the distribution of Ge($^1/_2$) equals one we would have to do the sum

$$^1/_2 + {}^1/_4 + {}^1/_8 + {}^1/_{16} + \cdots \tag{5.21}$$

which goes on for ever. However, infinite sums can be handled mathematically, and it is easy to prove that as we add in more and more of the terms in (5.21) the sum converges to one.

The geometric distributions are all J-shaped with a mode at $x=1$. The parameter p determines how rapidly the probabilities decrease as x increases – see Figure 5.5.

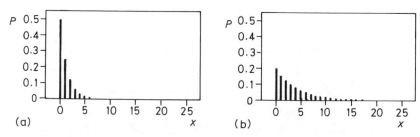

Figure 5.5. Distributions of (a) Ge(0.5), (b) Ge(0.2)

The negative binomial family

The random variable NB(n,p) is defined to have the negative binomial distribution with parameters n and p, whose probabilities are

$$P(\text{NB}(n,p)=x) = \binom{x+n-1}{n-1} p^n (1-p)^x ,$$

and where the possible values are $x = 0, 1, 2, \ldots$. The parameter p lies in $[0,1]$.

$$E(\text{NB}(n,p)) = n(1-p)/p .$$

$$var(\text{NB}(n,p)) = n(1-p)/p^2 .$$

The negative binomial family is a direct enlargement of the geometric family by the addition of a second parameter n. It is easy to see from their definitions that. for all values of p, $NB(1,p) \sim Ge(p)$.

Negative binomial distributions are in general unimodal and skewed to the right. The degree of skewness is greater as n decreases or as p increases. For $n=1$ we know that they are J-shaped. For larger n they will be J-shaped if p exceeds $1-n^{-1}$. The shapes are illustrated in Figure 5.6.

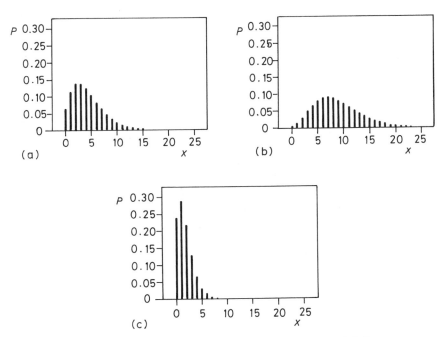

Figure 5.6. Distributions of (a) NB(3, 0.4), (b) NB(6, 0.4), (c) NB(4, 0.7)

The Poisson family

The random variable Po(m) is defined to have the Poisson distribution with mean m, whose probabilities are

$$P(\text{Po}(m)=x) = m^x e^{-m}/x!,$$

and whose possible values are $x = 0, 1, 2, \ldots$.

$$E(\text{Po}(m)) = m .$$

$$var(\text{Po}(m)) = m .$$

Notice that the mean and variance of Po(m) are both equal to m. For the binomial and hypergeometric distributions the variance is always less than the mean, whereas for geometric and negative binomial distributions the variance always exceeds the mean. The Poisson distributions occupy the middle ground.

Poisson distributions are unimodal and skewed to the right, with the skewness increasing as m decreases. For $m < 1$ they become J-shaped. See Figure 5.7, and Figure 5.1 for a Poisson distribution with large m.

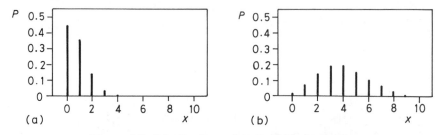

Figure 5.7. Distributions of (a) Po(0.8), (b) Po(4)

We have now completed our catalogue of standard distributions. With the aid of the descriptions given, in terms of mean, variance and shape, it is possible to fit measured distributions quickly and easily for a wide variety of problems. The example in Section 5.7 illustrated the technique but some further detail is needed. How exactly would one arrive at Po(52) rather than any other standard distribution? The answer lies in turning the loose measurements of location and dispersion into measurements of mean and standard deviation. This is achieved using some 'rules of thumb' which apply for all distributions which are unimodal and not too heavily skewed.

Measuring mean and standard deviation

Let X be a random variable whose distribution given H is unimodal and not greatly skewed. Then all usual measures of location will provide similar values, and any measurement of location can be considered as a measurement of $E(X|H)$. Furthermore, writing $m = E(X|H)$ and $s = sd(X|H)$,

$$P(m - s \leq X \leq m + s | H) \approx 0.65 .$$

$$P(m - 2s \leq X \leq m + 2s | H) \approx 0.95 .$$

Thus, for unimodal distributions the 'central' or 'typical' values are clear and unambiguous, so it is reasonable to interpret any measurement of such values as a measurement of the mean. The standard deviation is measured by considering that the range of values within one standard deviation of the mean has a total probability of about two-thirds. Alternatively, values within two standard deviations of the mean have a total probability of about 0.95. The physicist in Section 5.7, for instance, gave a typical value of 52. Since his distribution was also supposed to be unimodal and only slightly skewed, we assign $E_1(X|H) = 52$. The ranges 45 to 60 and 40 to 65, if interpreted as approximately one and two standard deviation bands around the mean, give a rough measurement of $sd_1(X|H) = 7$. Hence $var_1(X|H) = 49$. This is within measurement error of the mean, and an unbounded distribution seems realistic for this problem, so Po(52) is a reasonable choice of measured distribution.

{...........

Example

A lecturer is interested in $S =$ the number of students who will enroll on a certain course. From her information H, including enrollments in recent years, she estimates $S = 30$ as a most likely value, but enrollment is very variable. She only gives a probability of two-thirds to the range 20 to 45. Interpreting these measurements as implying

$$E_1(S|H) = 30, \quad sd_1(S|H) = 12, \quad var_1(S|H) = 144,$$

the variance greatly exceeds the mean. This indicates a negative binomial distribution. Equating

$$n(1-p)/p = 30, \qquad n(1-p)/p^2 = 144,$$

we must solve to give values for n and p. First

$$\frac{30}{144} = \frac{n(1-p)/p}{n(1-p)/p^2} = p,$$

so we set $p = 0.21$. Then

$$n \times 0.79/0.21 = 30 \qquad \therefore n = 30 \times 0.21/0.79 = 7.97.$$

The resulting measured distribution is therefore

$$S|H \; _2{\sim}\, NB(8, 0.21),$$

with

$$E_2(S|H) = 8 \times 0.79/0.21 = 30.1,$$

$$sd_2(S|H) = \{E_2(S|H)/0.21\}^{\frac{1}{2}} = 11.97.$$

The distribution is shown in Figure 5.8.

...........}

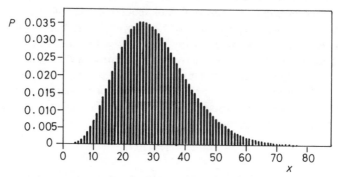

Figure 5.8. Measured distribution of $S=$ student enrollment, NB$(8, 0.21)$

Exercises 5(d)

1. Summarize, in terms of shape, location and dispersion, the biologist's distribution of Y given H obtained in Exercise 5(b)2.

2. Prove that equation (5.18) also implies $X|H \sim X|H \wedge (\neg E)$. Prove also that this independence relation has the usual symmetry property. That is, equation (5.18) asserts that X is independent of E: prove that E is independent of X by showing that for all possible values x of X,

$$P(E|H) = P(E|H \wedge (X{=}x)) .$$

3. It is a standard result in mathematics that

$$e^t = \sum_{n=0}^{\infty} t^n/n! ,$$

for any t. Use this to show that the sum of the probabilities $P(\text{Po}(m){=}x)$ for the Poisson random variable Po(m) is one. Prove also that $E(\text{Po}(m))=m$.

4. Consider your country's balance of payments figures (or any similar economic indicator) for the next twelve months. Let S be the number of months in which there is a balance of payments surplus. For each possible value $s=0,1,\ldots,12$ measure your probability $P(S{=}s|H)$ by direct measurement. Calculate the mean of your distribution. Calculate also the distribution of Bi$(12,p)$, with p chosen to give this distribution the same mean as your measured distribution. Compare the two distributions.

5. A mouse is placed in a circular box with six levers arranged around the edge. Pressing the correct lever will give the mouse a reward; pressing any other lever produces no result. The mouse presses levers at random until it receives the reward. 'At random' here means that each time it chooses a lever all six levers have an equal probability of being pressed, independently of earlier attempts.

Let X be the number of levers it presses before the mouse obtains its reward, and let $Y = X - 1$ (the number of wrong choices). Show that $Y|H \sim \text{Ge}(1/6)$.

Now suppose that after choosing the first lever at random (i.e. equal probabilities), the mouse then presses the levers in rotation, moving clockwise around the box. Show that $X|H \sim \text{U}(6)$. Which method gives the mouse the smaller expected number of attempts?

6. Using standard distributions, measure your distributions for the following random variables.

$S \equiv$ the number of symphonies composed by Dvořák.

$D \equiv$ the distance from Moscow to Leningrad.

$C \equiv$ the atomic number of caesium.

7. You buy a ticket for a lottery, in which just one winning ticket will be drawn. Let the random variable N be the number of tickets bought by other people. Let W be the proposition that yours will be the winning ticket. Your measured distribution for N is $\text{Po}(100)$. Elaborate your probability for W.

6

Distribution theory

This chapter develops a number of useful theoretical results which fall under the general heading of *distribution theory*. Distribution theory is the study of distributions, their summaries and relationships between them. The first two sections develop theory concerning the families of standard distributions introduced in Section 5.8. Section 6.1 is concerned with ways of deriving the standard distributions as objective distributions in appropriate problems, and Section 6.2 with relationships obtained by combining two random variables.

From a theoretical viewpoint the most useful summaries are those based on expectations. This is because of the variety of simple theorems relating expectations of functions of random variables. These constitute a part of distribution theory known as expectation theory. Section 6.3 deals with transformations of a single random variable. Transformations of two or more random variables are considered in Section 6.4 and the theorems of both sections are applied to give a collection of results about variances.

Random variables which are not independent are *associated* in some way. The remaining three sections deal with various aspects of association. The ideas of covariance and correlation are defined in Section 6.5, as ways of measuring degree of association. The regression function, defined in Section 6.6, is used to describe the nature of an association. It is defined in terms of conditional expectations, and Section 6.6 also presents some powerful theorems about conditional expectations generally. These are used in Section 6.7 to connect correlation and regression.

6.1 Deriving standard distributions

The study of distributions, their summaries and relationships between them, is known as distribution theory. Many of the results in Section 5.8 are basic distribution theory. In this chapter we present a little more theory from this field, partly to illustrate the techniques involved, but also because we will find some of the results useful in their own right. In this section we consider how various standard distributions may be characterized through appropriate derivations.

We have seen how standard distributions can be used to simplify the measurement of distributions. We have also seen, but not drawn attention to the fact, that they may arise as objective distributions. For instance, in Section 5.4 we considered the capture-recapture technique for learning about N = the number of fish in a lake. We noted that the distribution of X = second catch size, given $N = n$, was objectively determined by equation (5.12). This could be expressed more compactly as

$$X \mid H \wedge (N=n) \sim \mathrm{Hy}(n,6,10) .$$

There is no subscript before the $\sim$ because this is based on essentially objective recognition that when the number of fish is known, the second catch is simply a deal of ten from $(6, n-6)$.

The binomial and hypergeometric distributions arise in this objective way from sets of trials or deals, respectively. The other families defined in Section 5.8 also have derivations of this kind. The simplest is the geometric, which is the distribution of the number of failures before the first success in a series of trials. Since the random variable is clearly unbounded, it is necessary for the series of trials to be infinite in length.

Derivation of geometric distribution
In an infinite series of binary trials given H, with probability p of success in each trial, let X be the number of failures before the first success. Then

$$X \mid H \sim \mathrm{Ge}(p) .$$

{............

Proof
Let $E_i \equiv$ 'success at the i-th trial', then

$$'X=x' = E_{x+1} \wedge \bigwedge_{i=1}^{x} (\neg E_i) .$$

From the independence of trials, it follows immediately that

$$P(X=x \mid H) = p \, (1-p)^x .$$

............}

We have seen that the geometric distributions are a subfamily of the negative binomial distributions, with $\mathrm{Ge}(p) \sim \mathrm{NB}(1,p)$. This relationship appears in the derivation of the negative binomial distribution as the number of failures before the n-th success in a series of trials.

Derivation of negative binomial distribution
In an infinite series of binary trials given H, with probability p of success in each trial, let X be the number of failures before the n-th success. Then

$$X \mid H \sim \mathrm{NB}(n,p) .$$

{...........}

Proof

Let $E_i \equiv$ 'success at i-th trial'. There are many sequences of successes and failures in the first $n+x$ trials which correspond to '$X=x$'. For instance, one is

$$\{\bigwedge_{i=1}^{x} (\neg E_i)\} \wedge \{\bigwedge_{j=1}^{n} E_{x+j}\} .$$

In all such sequences there are n successes and x failures, so each sequence has probability $p^n (1-p)^x$ given H. It remains to count the number of such sequences. Each must end in a success, and otherwise any arrangement of $n-1$ successes in the first $x+n-1$ trials is possible. Therefore there are $\binom{x+n-1}{n-1}$ such sequences and from the sum theorem

$$P(X=x|H) = \binom{x+n-1}{n-1} p^n (1-p)^x .$$

{...........}

The derivation of the Poisson distributions is through a limiting process, in the same spirit as the demonstration in Section 4.11 that the distribution of $Bi(n,p)$ is obtained from the distribution of $Hy(N,R,n)$ if we let $N \to \infty$ and $R \to \infty$ such that $R/N=p$ is fixed.

Derivation of Poisson distribution
Let $X|H \sim Bi(n,p)$. If $n \to \infty$ and $p \to 0$ such that $np=m$ is fixed, then the distribution of X given H tends to that of $Po(m)$.

{...........}

Proof

$$P(X=x|H) = \binom{n}{x} p^x (1-p)^{n-x}$$

$$= \frac{n(n-1)\cdots(n-x)}{x!} \left(\frac{p}{1-p}\right)^x (1-p)^n$$

$$= \frac{m(m-p)\cdots(m-xp)}{x!(1-p)^x} \left(1-\frac{m}{n}\right)^n .$$

Now consider what happens to each part of this expression as we let $n \to \infty$, $p \to 0$ with $np=m$. First

$$m(m-p)\cdots(m-xp) \to m^x$$

because each of the x terms is tending to m as $p \to 0$. The $x!$ remains unchanged, but $(1-p)^x \to 1$. Last, a standard result of mathematics is that

$$(1-m/n)^n \to e^{-m}$$

as $n \rightarrow \infty$. Putting these limits together gives

$$P(X=x|H) \rightarrow m^x e^{-m}/x! .$$

..........}

In practice there are many situations for which this derivation makes the Poisson distribution appropriate. For example, let X be the number of calls received in one minute at a telephone switchboard. We could argue as follows. Let X_1 be the number of ten-second intervals in which at least one call is received, out of the six ten-second intervals making up one minute. If p_1 is the probability of at least one call in any ten-second interval, $X_1|H \sim \text{Bi}(6, p_1)$. Now let X_2 be the number of one-second intervals (out of the sixty making up a minute) with at least one call. Similarly we can say $X_2|H \sim \text{Bi}(60, p_2)$, where p_2 is much smaller than p_1. We could go on to let X_3 be the number of quarter-second intervals with at least one call, and so on. As the individual intervals get smaller, various things happen. First, the chance of more than one call arriving in any tiny interval becomes so minute that the number of intervals with at least one call becomes the same as our original random variable X. Second, the number of intervals becomes very large (tends to infinity). Third, the probability of even one call in an interval becomes very small (tends to zero). We can therefore think of X as the number of successes in a series of trials, where the number of trials is large but the chance of success in any trial is tiny.

Often we can think of events happening randomly and unconnectedly in time – telephone calls arriving at a switchboard, radio-active particles emitted from an isotope, cars passing along a road, leaves falling from a tree – where we are interested in how many events occur in a fixed time interval. Such random variables can be argued to have Poisson distributions. The main criterion for a Poisson argument to hold is independence between successive intervals. There should be no tendency for the events to clump together or for them to become regularly spaced. Using the example of cars passing a particular spot on a road, if this point is soon after a junction with traffic lights then there will be strong clumping of the events. Many cars may pass in one interval and none in the next. In this situation Your distribution for the number passing in a given interval would have a higher variance than the Poisson distribution would give. On the other hand, on a very busy, open stretch of road cars will tend to be fairly equally spaced, each driving a short distance behind the one in front. In this case we require a smaller variance than we obtain with the Poisson distribution.

The Poisson process

Suppose that events of a certain kind are occurring randomly in time, and You have no knowledge to suggest that they will tend either to clump together or to separate into regular spacing. Let X be the number of such events occurring in a fixed time interval. Then

$$X|H \sim \text{Po}(m),$$

where $m = E(X|H)$ is the average number of events occurring in time intervals of that length.

6.2 Combining distributions

Another group of results in distribution theory concern combinations of two or more standard distributions in joint distributions. Thus, if the distribution of X given H is a standard distribution, and if for each x the distribution of Y given $H \wedge (X=x)$ is another standard distribution, then by applying Bayes' theorem and extending the argument we may obtain other standard distributions for $Y|H$ and for $X|H \wedge (Y=y)$. Indeed, some families of standard distributions are characterized as being derived in this way. We consider one example of this kind of result here.

The binomial-Poisson combination theorem

Let X and Y be two random variables such that

$$X|H \sim \text{Po}(m), \qquad Y|H \wedge (X=x) \sim \text{Bi}(x,p).$$

Then

$$Y|H \sim \text{Po}(mp),$$

$$X|H \wedge (Y=y) \sim y + \text{Po}(m(1-p)). \tag{6.1}$$

Notice that on the right hand side of (6.1) appears a random variable $y + \text{Po}(m(1-p))$. This is simply the $\text{Po}(m(1-p))$ random variable with the constant y added to its value. Therefore the possible values are $y+z$, for $z = 0, 1, 2,$ ... and the probability that it takes the value $y+z$ is the same as the probability that $\text{Po}(m(1-p))$ equals z.

{...........

Proof

The joint distribution is

$$P((X=x)\wedge(Y=y)|H) = P(X=x|H)P(Y=y|H\wedge(X=x))$$

$$= (m^x e^{-m}/x!)\binom{x}{y}p^y(1-p)^{x-y}, \qquad (6.2)$$

for $x = 0, 1, 2, \ldots$, and $y = 0, 1, \ldots, x$. That is, the possible values of x and y are all pairs of non-negative integers such that $y \leq x$. To find the marginal distribution of Y we obtain $P(Y=y|H)$ by summing (6.2) over all possible x values. In this case the possible values of X when $Y=y$ are governed by the condition that $y \leq x$.

$$P(Y=y|H) = \sum_{x=y}^{\infty} P((X=x)\wedge(Y=y)|H)$$

$$= \frac{m^y p^y e^{-1}}{y!} \sum_{x=y}^{\infty} \frac{(m(1-p))^{x-y}}{(x-y)!}.$$

The sum is $\exp\{m(1-p)\}$ and we have

$$P(Y=y|H) = (mp)^y \exp(-mp)/y!. \qquad (6.3)$$

Next

$$P(X=x|H\wedge(Y=y)) = P((X=x)\wedge(Y=y)|H)/P(Y=y|H).$$

Dividing (6.2) by (6.3) gives

$$P(X=x|H\wedge(Y=y)) = (m(1-p))^{x-y}\exp\{-m(1-p)\}/(x-y)!,$$

which agrees with (6.1).

...........}

An intuitive explanation of the theorem is as follows. Think of X as the number of events occurring randomly in a time interval. These events are either successes or failures, with probability p of success, and the number of successes is Y. Therefore successes are also a kind of event occurring randomly in time, so Y has a Poisson distribution. If m events are expected in the interval then a proportion mp are expected to be successes, and this is the mean of Y. Failures are also occurring randomly in time, and by a similar argument their number has the distribution of $Po(m(1-p))$. Adding the number of successes, y, gives the conditional distribution of X, the number of events, if we know $Y=y$.

{...........

Example

Let $A \equiv$ the number of accidents occurring next month at a certain road junction. You consider that accidents happen randomly in time, and decide to assign a Poisson distribution. In determining its mean You estimate that there will be about seven accidents per year. Scaling down to one month,

$$A|H_1 \sim \text{Po}(^7/_{12}) \, .$$

Now let $B \equiv$ the number of accidents involving pedestrians at the same junction next month. You consider that approximately one-third of all accidents involve pedestrians, and so You judge

$$B|(A=a) \wedge H_1 \sim \text{Bi}(a, ^1/_3) \, .$$

(This is not strictly realistic because the events are not independent. If You learn that four accidents have already occurred and all of them involved pedestrians, then Your probability for the next accident involving a pedestrian will surely increase to more than one-third. We cannot deal properly with such a case until Chapter 8.)

From the theorem we now have

$$B|H_2 \sim \text{Po}(^7/_{36}) \, , \quad A|(B=b) \wedge H_2 \sim b + \text{Po}(^7/_{18}) \, .$$

For instance, Your measured probability that no accidents will occur involving pedestrians is

$$P_2(B=0|H) = \exp(-^7/_{36}) = 0.82 \, .$$

...........}

Exercises 6(a)

1. Let $R|H \sim \text{Bi}(N,p)$ and $X|(R=r) \wedge H \sim \text{Hy}(N,r,n)$ for $r=0, 1, \ldots, N$. Prove that $X|H \sim \text{Bi}(n,p)$ and $R|(X=x) \wedge H \sim x + \text{Bi}(N-n,p)$.

2. Explain the results of Exercise 1 via the characterizations of binomial and hypergeometric distributions.

3. You are observing birds visiting a bird table. You think of their arrivals as a Poisson process, and if M is the number of arrivals in one minute Your distribution for M is Po(4). What distribution should You give for T = the number of arrivals in thirty seconds?

6.3 Basic theory of expectations

One reason for using the mean and standard deviation as measures of location and dispersion is that it is possible to develop some very neat theoretical results concerning expectations. We present some of these results here and in the next section.

First consider transformations of a random variable. For instance, if a is some constant then the random variable $Y = a + X$ is defined on the same partition as X, but takes the value $y = a + x$ when $X = x$. In general, if g is a function taking as its argument the possible values of X, we can define the transformed random variable $Y = g(X)$.

Transformations of a random variable

Let X be a random variable, and let g be a function giving a value $g(x)$ for each possible value x of X. Then the random variable $g(X)$ is defined such that if X takes the value x then $g(X)$ takes the value $g(x)$.

If every possible value x of X yields a different transformed value $g(x)$, then $g(X)$ is defined on the same partition as X. Such a transformation is called *one-to-one*.

If g is not a one-to-one transformation then $g(X)$ has fewer possible values than X. The proposition '$g(X) = y$' equals the disjunction of all the various propositions '$X = x$' for which $g(x) = y$. Whilst this is in general a complication when we deal with transformed random variables, the following theorem shows that expectations can be calculated by the same formula whether or not the transformation is one-to-one.

The expectation of a transformation theorem

Let X be a random variable, then for any information H, and any function g,

$$E\{g(X)|H\} = \sum_x g(x) P(X=x|H),\qquad(6.4)$$

where the sum is over all possible values x of X.

{...........

Proof

By definition,

$$E\{g(X)|H\} = \sum_y y\, P(g(X)=y|H),\qquad(6.5)$$

summing over the possible values y of $g(X)$. If G is one-to-one then the propositions '$X = x$' and '$g(X) = g(x)$' are identical, so we can simply replace y by $g(x)$ and sum over x, obtaining (6.4) immediately.

If g is not one-to-one then by the sum theorem

$$P(g(X)=y|H) = \sum_{g(x)=y} P(X=x|H),\qquad(6.6)$$

the sum being over all those x for which $g(x) = y$. Now if we use (6.6) in (6.5) we are taking each value y in turn, summing over those x for which

$g(x) = y$, and summing the result over y. The effect is of summing over all possible values x of X. Moreover, the term y in (6.5) is always equated to $g(x)$, and the result is (6.4) again.

..........}

{..........

Example

Let X have the distribution given H shown below.

x	1	2	3	4	5	6
$P(X=x|H)$	0.1	0.2	0.2	0.2	0.2	0.1

Consider the random variable $|X^3 - 9X^2 + 23X - 15| = g(X)$, where

$$g(x) = |x^3 - 9x^2 + 23x - 15|,$$

the bars $|\ ...\ |$ denoting absolute value as usual. Then $g(x) = 0$ when $x = 1, 3$ or 5, $g(x) = 3$ when $x = 2$ or 4 and $g(6) = 15$. The distribution of $g(X)$ is therefore

y	0	3	15
$P(g(X)=y|H)$	0.5	0.4	0.1

If we now compute $E\{g(X)|H\}$ direct from the above table we have

$$E\{g(X)|H\} = (0 \times 0.5) + (3 \times 0.4) + (15 \times 0.1) = 2.7.$$

Alternatively, using (6.4) gives

$$E\{g(X)|H\} = g(1) \times 0.1 + g(2) \times 0.2 + ... + g(6) \times 0.1$$

$$= (0 \times 0.1) + (3 \times 0.2) + (0 \times 0.2) + (3 \times 0.2) + (0 \times 0.2) + (15 \times 0.1)$$

$$= 0 \times (0.1 + 0.2 + 0.2) + 3 \times (0.2 + 0.2) + 15 \times 0.1 = 2.7.$$

This demonstrates how (6.4) works through equations (6.5) and (6.6).

..........}

The expectation of a transformation theorem is useful because it allows us to calculate $E\{g(X)|H\}$ without actually finding the distribution of $g(X)$ given H. We return now to the simple transformation $Y = a + X$. Since Y is just X shifted by an amount a, we would expect that a location measure of Y would be a similarly shifted location measure of X. This is true of the expectation, and the following theorem extends the result to the more general linear transformation $Y = a + bX$.

Expectation of a linear function
For any random variable X, any information H and any two constants a and b

$$E(a+bX|H) = a + b\,E(X|H).$$ (6.7)

{...........

Proof

$$E(a+bX|H) = \sum_x (a+bx)\,P(X=x|H)$$

$$= \sum_x a\,P(X=x|H) + \sum_x bx\,P(X=x|H)$$

$$= a \sum_x P(X=x|H) + b \sum_x x\,P(X=x|H)$$

..........}

Because of (6.7), the expectation is sometimes referred to as a 'linear operator'.

6.4 Further expectation theory

We will next prove a theorem which adds considerably to the importance of expectations. It deals with the random variable $Z=X+Y$, which is a transformation of two random variables into one. The idea is a simple extension of transformations of a single random variable. In general, let g be a function with two arguments. Then the random variable $g(X,Y)$ is defined such that when $X=x$ and $Y=y$ then $g(X,Y)=g(x,y)$ for all possible values x of X and y of Y. Functions of three or more random variables are defined in the obvious way.

Transformations of two or more random variables
Let $X_1, X_2, \ldots, X_n$ be random variables, and let g be a function giving a value $g(x_1, x_2, \ldots, x_n)$ for all combinations of possible values x_i of X_i ($i=1, 2, \ldots, n$). Then the transformed random variable $g(X_1, X_2, \ldots, X_n)$ is defined such that when X_i takes the value x_i, $i=1, 2, \ldots, n$, then $g(X_1, X_2, \ldots, X_n)$ takes the value $g(x_1, x_2, \ldots, x_n)$. Its expectation may be computed as

$$E\{g(X_1, X_2, \ldots, X_n)|H\}$$
$$= \sum_{x_1} \sum_{x_2} \cdots \sum_{x_n} g(x_1, x_2, \ldots, x_n)\,P(\wedge_{i=1}^n (X_i=x_i)|H),$$

the sums being over all possible values x_i of X_i.

The expectation result is proved in the same way as equation (6.4).

Expectation of a sum of random variables

For any two random variables X and Y, and any information H,

$$E(X+Y|H) = E(X|H) + E(Y|H). \tag{6.8}$$

{...........

Proof

$$\begin{aligned}
E(X+Y|H) &= \sum_x \sum_y (x+y) P((X=x) \wedge (Y=y)|H) \\
&= \sum_x \sum_y x \, P((X=x) \wedge (Y=y)|H) \\
&\quad + \sum_x \sum_y y \, P((X=x) \wedge (Y=y)|H) \\
&= \sum_x x \left[\sum_y P((X=x) \wedge (Y=y)|H) \right] \\
&\quad + \sum_y y \left[\sum_x P((X=x) \wedge (Y=y)|H) \right] \\
&= \sum_x x \, P(X=x|H) + \sum_y y \, P(Y=y|H)
\end{aligned}$$

..........}

As an example of this theorem, look at Section 5.5, where $Z=X+Y$. From Tables 5.13 and 5.15 we find $E(Z|H)=2.6$, $E(X|H)=1.73$ and $E(Y|H)=0.87$.

Notice that this theorem applies for any two random variables at all. They may themselves be functions, even sums, of other random variables. So an immediate corollary is that, for any random variables $X_1, X_2, \ldots, X_n$,

$$E(\Sigma_{i=1}^n X_i | H) = \sum_{i=1}^n E(X_i|H). \tag{6.9}$$

We can often use equations (6.8) and (6.9) even when the distribution of $X+Y$ or $\Sigma_{i=1}^n X_i$ is very much more complicated to calculate.

Products of random variables are more difficult. It is not in general true that the expectation of a product equals the product of expectations. We require an extra condition – independence.

Expectation of a product of random variables

Let X and Y be independent given H. Then

$$E(XY|H) = E(X|H)E(Y|H). \tag{6.10}$$

{...........
   Proof

$$E(XY|H) = \sum_x \sum_y xy \, P((X=x) \wedge (Y=y)|H)$$

$$= \sum_x \sum_y xy \, P(X=x|H) P(Y=x|H)$$

$$= \sum_x x \, P(X=x|H) \left[\sum_y y \, P(Y=y|H) \right]$$

$$= \sum_x x \, P(X=x|H) E(Y|H)$$

$$= E(X|H) E(Y|H) .$$

..........}

Again, this applies when X and Y are themselves functions of other random variables. Notice that it is simple to prove that if X and Y are independent then so are any two transformed random variables $g(X)$ and $h(Y)$.

An obvious extension of (6.10) is that if $X_1, X_2, \ldots, X_n$ are mutually independent given H then

$$E(\Pi_{i=1}^n X_i | H) = \prod_{i=1}^n E(X_i | H) .$$

We can now assemble all these ideas and theorems into a useful set of results about variances. The following theorem demonstrates the power of expectation theory.

The variance as an expectation

Let X be any random variable and H be any information. Let $m = E(X|H)$. Then

   (*a*) $var(X|H) = E\{(X-m)^2|H\}$,   (6.11)

   (*b*) $var(X|H) = E(X^2|H) - m^2$,   (6.12)

   (*c*) $var(X|H) = E\{X(X-1)|H\} + m - m^2$ (6.13)

             $= E\{X(X+1)|H\} - m - m^2$.   (6.14)

If a and b are any two constants then

   (*d*) $var(a+bX|H) = b^2 var(X|H)$.   (6.15)

Let X and Y be independent given H, then

   (*e*) $var(X+Y|H) = var(X|H) + var(Y|H)$.   (6.16)

Distribution Theory

{.............

Proof

(a) follows immediately from (6.4) and the definition (5.14). (b) follows from (a) via (6.5) and (6.7). Thus

$$var(X|H) = E(X^2 - 2mX + m^2|H)$$
$$= E(X^2|H) + E(-2mX + m^2|H)$$
$$= E(X^2|H) - 2mE(X|H) + m^2$$
$$= E(X^2|H) - 2m^2 + m^2.$$

(c) follows from (b) and (6.8) since

$$E(X(X-1)|H) = E(X^2 - X|H) = E(X^2|H) - m,$$

and similarly for $E(X(X+1)|H)$. (d) follows from (b) using

$$var(a + bX|H) = E\{(a + bX)^2|H\} - \{E(a + bX|H)\}^2.$$

Expanding the first term gives

$$E\{(a + bX)^2|H\} = a^2 + 2abm + b^2 E(X^2|H).$$

Expanding the second term gives

$$(a + bm)^2 = a^2 + 2abm + b^2m^2.$$

Subtracting and using (b) again proves (d). (e) also follows from (b):

$$var(X + Y|H) = E\{(X + Y)^2|H\} - \{E(X + Y|H)\}^2.$$

The first term becomes

$$E(X^2 + 2XY + Y^2|H) = E(X^2|H) + 2E(X|H)E(Y|H) + E(Y^2|H)$$

using (6.10). The second term is, from (6.8)

$$\{E(X|H) + E(Y|H)\}^2$$
$$= \{E(X|H)\}^2 + 2E(X|H)E(Y|H) + \{E(Y|H)\}^2.$$

.............}

Equations (6.12), (6.13) and (6.14) may be employed to derive all the variances given for standard distributions in Section 5.8.

{.............

Example

Consider $var(\text{Bi}(n,p))$. Using a similar technique to that used to prove $E(\text{Bi}(N,p)) = np$ we have

$$E\{\text{Bi}(n,p)(\text{Bi}(n,p) - 1)\}$$

$$= \sum_{x=0}^{n} x(x-1)\binom{n}{x}p^x (1-p)^{n-x}$$

$$= \sum_{x=2}^{n} \frac{n!}{(x-2)!\,(n-x)!} p^x (1-p)^{n-x}$$

$$= n(n-1)p^2 \sum_{x^*=0}^{n^*} \binom{n^*}{x^*}p^{x^*} (1-p)^{n^*-x^*}$$

$$= n(n-1)p^2 ,$$

where $x^*=x-2$ and $n^*=n-2$. From (6.13),

$$var\,(\mathrm{Bi}(n,p)) = n(n-1)p^2 + np - n^2p^2$$

$$= np - np^2 = np\,(1-p) .$$

..........}

Exercises 6(b)

1. Consider the random variables T, L, S and P of Exercise 5(a)1 and their expectations found in Exercise 5(c)4. Why do we find that $E(T|H)=5-E(S|H)$ and $E(L|H)=E(S|H)+2E(T|H)$, but $E(P|H)\neq 2E(T|H)/E(L|H)$?

2. Four dice are tossed. Let F be the total score. Show that the expectation of F is 14 and obtain its variance. (Do not attempt to calculate its distribution!)

3. Consider the random variables M and T in Exercise 6(a)3 and define $S\equiv M/2$. Show that $E(T|H)=E(S|H)$ but $var\,(T|H)\neq var\,(S|H)$. Why?

4. Let $E_1, E_2, \ldots, E_n$ be any propositions and define corresponding random variables $X_1, X_2, \ldots, X_n$ by $X_i=1$ if E_i is true and $X_i=0$ if E_i is false. (X_i is called the *indicator random variable* of E_i.) Let $p_i=P(E_i|H)$. Consider $Y\equiv\sum_{i=1}^{n} X_i=$ the number of E_is which are true, and prove

$$E(Y|H) = \sum_{i=1}^{n} p_i .$$

If the E_is are independent given H, prove that

$$var\,(Y|H) = \sum_{i=1}^{n} p_i(1-p_i) .$$

Use these results to obtain the mean and variance of $\mathrm{Bi}(n,p)$.

5. Use the indicator random variables of Exercise 4 to show that $E(\mathrm{Hy}(N,R,n))=n\,R/N$.

6. Let X be any random variable and let $d(a)\equiv E\{(X-a)^2|H\}$ for all a. Show that the value of a which minimizes $d(a)$ is $E(X|H)$.

7. Show that

$$E\left[\text{Po}(m)\left\{\text{Po}(m)-1\right\}\right]=m^2,$$

and hence that $var\left(\text{Po}(m)\right)=m$.

6.5 Covariance and correlation

Two random variables which are not independent may be said to be *associated* in some way. In this section we consider some ways of looking at association.

One simple approach is based on the fact that (6.16) is not generally true if X and Y are associated. Following through the proof of (6.16) we find the following more general expression.

$$var\left(X+Y|H\right)=var\left(X|H\right)+var\left(Y|H\right)+2\,cov\left(X,Y|H\right),\qquad(6.17)$$

where

$$cov\left(X,Y|H\right)=E\left(XY|H\right)-E\left(X|H\right)E\left(Y|H\right).\qquad(6.18)$$

Equation (6.17) is true for any X and Y, and for any information H. From (6.18) and (6.10), if X and Y are independent then $cov\left(X,Y|H\right)=0$, so (6.17) reduces to (6.16) in this case. Otherwise we could consider the magnitude of $cov\left(X,Y|H\right)$ as indicating the degree of association. Covariances are based on expectations and enjoy very similar properties to variances.

Covariance
The covariance of random variables X and Y on information H is denoted by $cov\left(X,Y|H\right)$ and defined as

$$cov\left(X,Y|H\right)=E\left\{\left(X-E\left(X|H\right)\right)\left(Y-E\left(Y|H\right)\right)|H\right\}\qquad(6.19)$$

$$=E\left(XY|H\right)-E\left(X|H\right)E\left(Y|H\right).$$

For any constants $a_1,a_2,b_1,b_2,$

$$cov\left(a_1+b_1X,a_2+b_2Y|H\right)=b_1b_2\,cov\left(X,Y|H\right).\qquad(6.20)$$

For any four random variables W,X,Y,Z and information H,

$$cov\left(W+X,Y+Z|H\right)$$

$$=cov\left(W,Y|H\right)+cov\left(W,Z|H\right)$$

$$+cov\left(X,Y|H\right)+cov\left(X,Z|H\right).\qquad(6.21)$$

{...........}

Proof

We can first derive (6.18) from (6.19), to show that they are equivalent. To simplify notation, let $m = E(X|H)$, $q = E(Y|H)$. Then (6.19) is

$$cov(X, Y|H) = E\{(X-m)(Y-q)|H\}$$
$$= E(XY - mY - qX + mq|H)$$
$$= E(XY|H) - mq - qm + mq,$$

which reduces to (6.18). Results (6.20) and (6.21) are easy to prove using (6.18) and the techniques of Section 6.4.

...........}

The connection between variance and covariance becomes clearer still if we replace Y by X in (6.18) or (6.19), and we find

$$cov(X, X|H) = var(X|H). \tag{6.22}$$

In other words, the variance of X can be seen as its covariance with itself, and covariance is a generalization of variance. Letting $a_1 = a_2, b_1 = b_2, Y = X$ in (6.20) reproduces (6.15), and letting $W = Y$, $X = Z$ in (6.21) reproduces (6.17).

Covariance is not an ideal measure of association because, as seen in (6.22), it is connected with dispersion. Increasing the dispersion of X and Y by rescaling them, as in (6.20), with factors b_1 and b_2 greater than one increases covariance. Rescaling with factors less than one reduces both variance and covariance. Therefore as a measure of association we use a scaled version of covariance, called the *correlation coefficient*, which is not sensitive to dispersion.

The correlation coefficient

The correlation coefficient between X and Y given H is denoted by $corr(X, Y|H)$ and defined as

$$corr(X, Y|H) \equiv cov(X, Y|H) / \{sd(X|H) sd(Y|H)\}. \tag{6.23}$$

For any X, Y, H,

$$-1 \leq corr(X, Y|H) \leq 1. \tag{6.24}$$

For any constants a_1, a_2, b_1, b_2,

$$corr(a_1 + b_1 X, a_2 + b_2 Y|H) = corr(X, Y|H). \tag{6.25}$$

If $corr(X, Y|H) = 0$ then X and Y are said to be *uncorrelated* given H. If X and Y are independent given H then they are also uncorrelated given H.

{...........

Proof

To prove (6.24) we prove that the square of $corr(X,Y|H)$ is less than or equal to one. Consider

$$var(X+bY|H) = var(X|H) + b^2var(Y|H) + 2b\,cov(X,Y|H),$$

using (6.17), (6.15) and (6.20). This is true for any constant b, and we choose

$$b = -cov(X,Y|H)/var(Y|H). \tag{6.27}$$

Substituting (6.27) into (6.26) we find

$$var(X+bY|H)$$

$$= var(X|H) - \{cov(X,Y|H)\}^2/var(Y|H). \tag{6.28}$$

Now since this is a variance, the right hand side of (6.28) must be greater than or equal to zero. The fact that $\{corr(X,Y|H)\}^2 \le 1$ then follows.

Equation (6.25) is a simple consequence of (6.15) and (6.20). The fact that independence implies zero correlation is also immediate.

...........}

X and Y may be uncorrelated without being independent but this is rare in practice. Generally, the further $corr(X,Y|H)$ is from zero, either positive or negative, the 'less independent' they are. However, we need some more concepts before we can make the notion of 'less independent' clear.

{...........

Example

To illustrate the computations, consider the joint distribution of X and Y in Table 5.15. First we compute the (measured) covariance, using (6.18).

$$E_1(XY|H) = (0\times0\times0.10) + (0\times1\times0.07) + (0\times2\times0.01) + (1\times0\times0.13)$$

$$+ (1\times1\times0.11) + ... + (4\times2\times0.05)$$

$$= 1.85.$$

$$\therefore cov_1(X,Y|H) = 1.85 - 1.73\times0.87 = 0.345.$$

The variances of X and Y given H are found to be 1.48 and 0.57 respectively. Hence

$$corr_1(X,Y|H) = 0.345/(1.48\times0.57)^{1/2} = 0.38,$$

reflecting a moderate degree of correlation between X and Y given H.

...........}

Exercises 6(c)

1. As part of a study into the epidemiology of a disease, the following joint distribution is proposed for N and M, the numbers of cases in a certain area in two successive months.

n	0	1	2	3	4
m					
0	0.112	0.089	0.056	0.022	0.006
1	0.089	0.084	0.067	0.039	0.015
2	0.056	0.067	0.058	0.034	0.019
3	0.022	0.039	0.034	0.025	0.011
4	0.006	0.015	0.019	0.011	0.005

Calculate the covariance and correlation coefficient between N and M for this distribution.

2. Calculate the covariance and correlation between X and Y in Exercise 5(b)2.

3. Let X and Y have the same marginal distributions as in Exercise 5(a)3, but suppose that they are not independent. If their correlation coefficient is positive, would the variance of $Z=X+Y$ be larger or smaller than the value found in Exercise 5(c)2?

4. Let X have the distribution shown below and let $Y=X^2$. Show that X and Y are not independent but that they are uncorrelated.

x	−2	−1	0	1	2
$P(X=x|H)$	0.1	0.2	0.4	0.2	0.1

[More mathematically-inclined readers might prove that this is always true if the distribution of X is symmetric and centred at zero.].

6.6 Conditional expectations

We know that one way of thinking about independence between X and Y is that if, in addition to H, You learn the value of one variable then Your probabilities for the other are unchanged. Learning about one conveys no information about the other. Association, then, is the opposite: learning about X conveys information to You about Y, and vice versa. Naturally, there are many ways of describing association in these terms. We will consider summaries based on expectations.

The *regression function* of X on Y given H is defined to be $E(X|(Y=y)\wedge H)$. Notice that this expectation takes a different value for each possible value y of Y. As y varies we consider $E(X|(Y=y)\wedge H)$ as a function of y. If X and Y are independent given H

$$E(X|(Y=y)\wedge H) = E(X|H) .$$

This is trivial to prove. Otherwise, when they are associated, the regression function varies with y. For instance, the table in the first example given in Section 5.6 shows the (measured) regression function of N on X given H in the capture-recapture problem. As x increases from 0 to 6, $E_1(N|(X=x)\wedge H)$ decreases steadily from 24.4 to 19.0. The two variables are clearly associated, and learning about X conveys information about N.

Regression function

The regression function of X on Y given H is defined to be $E(X|(Y=y)\wedge H)$ considered as a function of the possible values y of Y. If X and Y are independent given H then the regression function of X on Y given H is constant and equal to $E(X|H)$.

Similarly, the regression function of Y on X given H is $E(Y|(X=x)\wedge H)$, a function of x. The notion of regression, unlike independence or correlation, is not symmetric. There is no special relationship between the regression functions of X on Y and Y on X.

For some values of y, $E(X|(Y=y)\wedge H)$ could be far from $E(X|H)$ and for other values of y they could be similar. For some y, $E(X|(Y=y)\wedge H)$ might be greater than $E(X|H)$ and for others it might be smaller. However, there is an important result which says that if the regression function exceeds $E(X|H)$ for some y then this must be balanced by it being less than $E(X|H)$ for some other values. The basis of this result is to consider the random variable obtained by transforming Y by the regression function of X on Y given H.

The expectation of an expectation theorem

For two random variables X and Y, and information H, let $E(X|Y\wedge H)$ be the random variable $g(Y)$ defined by transforming Y with the regression function

$$g(y) = E(X|(Y=y)\wedge H) . \tag{6.29}$$

Then

$$E(X|H) = E\{E(X|Y\wedge H)|H\} . \tag{6.30}$$

{ ....

Proof

$$g(y) = \sum_x x\, P(X=x|(Y=y)\wedge H) ,$$

the sum being over all possible values x of X. Therefore,

$$E\{E(X|Y \wedge H)|H\} = E\{g(Y)|H\} = \sum_y g(y)P(Y=y|H)$$

$$= \sum_y \sum_x x\,P(X=x|(Y=y) \wedge H)P(Y=y|H)$$

$$= \sum_y \sum_x x\,P((X=x) \wedge (Y=y)|H).$$

The sum is over all possible pairs x and y. Summing over y first gives

$$\sum_x \{\sum_y P((X=x) \wedge (Y=y)|H)\} = \sum_x x\,P(X=x|H).$$
...........}

{...........

Example

In the capture-recapture problem we computed the regression function of N on X given H. Letting $g(0)=24.4$, $g(1)=23.4, \ldots, g(6)=19.0$, we find (with the distribution of X given in Table 5.10)

$$E_1\{E_1(N|X \wedge H)|H\} = E_1\{g(X)|H\}$$

$$= (0.014 \times 24.4) + (0.101 \times 23.4) + ... + (0.005 \times 19.0) = 21.47.$$

We earlier computed $E_1(N|H)=21.66$. The small difference is caused by rounding the figures in various calculations.
...........}

Equation (6.30) is closely related to extending the argument. If it is easier to think about probabilities for X when the value of Y is known, then the regression function of X on Y is readily determined, from which the marginal mean of X may be derived using (6.30). The following theorem allows the marginal variance of X to be calculated by the same extension.

The conditional variance theorem

For random variables X and Y define $E(X|Y \wedge H)$ as in (6.29) and define $var(X|Y \wedge H)$ to be the random variable $h(Y)$ obtained from the transformation

$$h(y) = var(X|(Y=y) \wedge H).$$

Then

$$var(X|H) = E\{var(X|Y \wedge H)|H\}$$

$$+ var\{E(X|Y \wedge H)|H\}. \tag{6.31}$$

{...........

Proof

Consider the two terms on the right hand side of (6.31). The first is

$$E\{var(X|Y \wedge H)|H\} = E[E(X^2|Y \wedge H) - \{E(X|Y \wedge H)\}^2|H]$$

$$= E\{E(X^2|Y \wedge H)|H\} - E\{g(Y)^2|H\}.$$

From (6.30) applied to the random variables X^2 and Y this reduces further.

$$E\{var(X|Y \wedge H)|H\} = E(X^2|H) - E\{g(Y)^2|H\}. \tag{6.32}$$

Next

$$var\{E(X|Y \wedge H)|H\} = E[\{E(X|Y \wedge H)\}^2|H]$$

$$- [E\{X|Y \wedge H)|H\}]^2$$

$$= E\{g(Y)^2|H\} - \{E(X|H)\}^2,$$

using (6.30). Combining this with (6.32) gives (6.31).

...........}

6.7 Linear regression functions

Equation (6.31) can be made the basis of a measure of association. For, observe that the second term, being a variance, is greater than or equal to zero. It only equals zero if the random variable $E(X|Y \wedge H)$ takes only one possible value, i.e. if $E(X|(Y=y) \wedge H)$ is a constant. Therefore

$$var(X|H) \geq E\{var(X|Y \wedge H)|H\}, \tag{6.33}$$

with equality if X and Y are independent given H. We alluded to this result in Section 5.6, when we noted that the posterior variance of N given $(X=x) \wedge H$ was always less than its prior variance. Now, it is not always true that posterior variance is less than or equal to prior variance, but equation (6.33) says that the expected posterior variance cannot exceed the prior variance. Therefore, gathering new information is 'on average' rewarded by a reduction in variance.

The difference between $var(X|H)$ and $E\{var(X|Y \wedge H)|H\}$ represents the degree to which Y is informative about X, and in this sense measures how associated X and Y are. The difference itself, like the covariance, is sensitive to dispersion, and so we scale it accordingly.

An information measure

The quadratic information provided by Y about X given H is denoted by $I(X,Y|H)$ and defined as

$$I(X,Y|H) = 1 - E\{ var(X|Y \wedge H)|H \}/var(X|H) \qquad (6.34)$$

$$= var\{ E(X|Y \wedge H)|H \}/var(X|H).$$

{..........

Example

The prior and posterior variances of N in the capture-recapture example are shown in the examples in Section 5.6. First, $var_1(N|H) = 3.02^2 = 9.12$, then using Table 5.10,

$$E_1\{ var(N|X \wedge H)|H \} = 2.82^2 0.014 + 2.88^2 0.101 + ... +$$

$$+ 2.58^2 0.005 = 8.04$$

$$\therefore I_1(N,X|H) = 1 - 8.04/9.12 = 0.12.$$

Referring also to the values of $E(N|(X=x) \wedge H)$ given in Section 5.6 we can verify equation (6.31) as follows.

$$E_1\big[\{ E_1(N|X \wedge H) \}^2 |H \big] = 24.4^2 0.014 + ... + 19.0^2 0.005 = 470.75.$$

Subtracting $\{ E_1(N|H) \}^2 = 21.672^2 = 469.16$ leaves $1.08 = 9.12 - 8.04$.

..........}

One disadvantage of $I(X,Y|H)$ is that it is not symmetric, i.e. $I(X,Y|H) \neq I(Y,X|H)$. In practice, the two quadratic information measures tend to be quite close together. The following theorem supplies the reason.

Linear regression

Let the regression of X on Y given H be linear. That is,

$$E(X|(Y=y) \wedge H) = a + by, \qquad (6.36)$$

where a and b are constants. Then

$$a = E(X|H) - E(Y|H)\{ cov(X,Y|H)/var(Y|H) \}, \qquad (6.37)$$

$$b = cov(X,Y|H)/var(Y|H) \qquad (6.38)$$

and

$$I(X,Y|H) = \{ corr(X,Y|H) \}^2. \qquad (6.39)$$

{...........

Proof

If (6.36) is true then

$$E(X|H) = E\{E(X|Y \wedge H)|H\}$$

$$= E(a+bY|H) = a + bE(Y|H). \tag{6.40}$$

Now let $Z = XY$ and consider

$$E(XY|H) = E(Z|H) = E\{E(Z|Y \wedge H)|H\}.$$

But

$$E(XY|(Y=y) \wedge H) = y\,E(X|(Y=y) \wedge H) = ay + by^2,$$

$$\therefore E(XY|H) = E(aY + bY^2|H) \tag{6.41}$$

$$= aE(Y|H) + bE(Y^2|H).$$

$$\therefore cov(X,Y|H) = E(XY|H) - E(X|H)E(Y|H)$$

$$= b\,var(Y|H),$$

using (6.40) and (6.41). Equation (6.38) is now immediate and (6.37) follows using (6.40). To prove (6.39) we take (6.36) and obtain

$$var\{E(X|Y \wedge H)|H\} = var(a + bY|H) = b^2 var(Y|H)$$

$$= \{cov(X,Y|H)\}^2/var(Y|H),$$

using (6.38). Using now (6.35) gives (6.39).
...........}

We can now see that if both regression functions, X on Y and Y on X, are close to straight lines then both information measures will be close to the square of the correlation coefficient, and hence close to each other. It often happens in practice that regression lines are reasonably straight.

{...........

Example

In the capture-recapture problem we found that the regression function of N on X given H was as shown in Figure 6.1. The points lie very near a straight line for $X=0$ to $X=5$, but curve away slightly at $X=6$. We found $I_1(N,X|H) = 0.12$ and the correlation coefficient could also be calculated form Tables 5.10 and 5.11. In this particular problem there is a neater way. Using the objective distributions of X given $(N=n) \wedge H$, i.e.

$$X|(N=n) \wedge H \sim Hy(n,10,6),$$

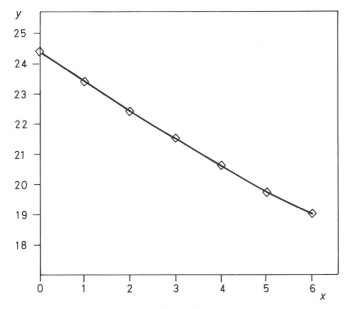

Figure 6.1. Regression of N on X, capture-recapture example

we know that

$$E(X|(N=n)\wedge H) = 60/n \ .$$

Therefore, using the expectation of an expectation theorem,

$$E(NX|H) = E\{ E(NX|N \wedge H)|H \}$$
$$= E(N \times 60/N|H) = 60 \ .$$

From this we can quickly find the covariance, and thence the correlation coefficient. We find $\{corr(N,X|H)\}^2 = 0.345^2 = 0.12$. To within the accuracy of our calculations at least, the quadratic information measure is actually equal to the square of the correlation coefficient.

...........}

The linear regression theorem allows us a clear way of interpreting the correlation coefficient, assuming that neither regression function is dramatically curved. The further the correlation coefficient is from zero the greater is its square, which represents approximately the amount of information that either variable gives about the other. The sign of the correlation coefficient indicates the direction of slope of the regression functions. For if $corr(X,Y|H) > 0$ then the regression function of X on Y given H generally increases with y and the regression function of Y on X also generally increases with x. Whereas, if $corr(X,Y|H) < 0$ then both regressions tend to be decreasing functions.

An appreciation of association, through the notions of covariance, correlation, regression and information, is valuable in studying the joint distributions of two or more random variables. It is equally important in measuring joint distributions. However, using these ideas in measuring distributions requires some experience, and we shall not pursue this approach further.

Exercises 6(d)

1. Consider the random variables X and Y in the binomial-Poisson combination theorem of Section 6.2. Write down the regression functions of X on Y and Y on X. Obtain the covariance between X and Y using (6.38). Derive the information measures $I(X,Y|H)$ and $I(Y,X|H)$, confirming that they are equal and that (6.39) holds.

2. Calculate the regression function of N on M for the joint distribution in Exercise 6(c)1. Obtain $I(N,M|H)$ and compare it with $corr(N,M|H)^2$.

3. Prove that for any random variables X, Y and Z, and information H,

$$cov(X,Y|H) = E\{cov(X,Y|Z \wedge H)|H\}$$

$$+ cov\{E(X|Z \wedge H), E(Y|Z \wedge H)|H\}.$$

4. A workshop has 14 machines. Because of staff illness and holidays they are not usually all in use. Let N = the number of machines in use on a given day. Your distribution of N has mean and variance

$$E(N|H) = 13, \qquad var(N|H) = 2.8.$$

Given $N = n$, let X_i be the number of breakdowns experienced by machine i, $i = 1, 2, \ldots, n$, on that day. Let $B = \sum_{i=1}^{n} X_i$ be the total number of breakdowns. Given $(N = n) \wedge H$ the X_is are mutually independent with Poisson distributions,

$$X_i|(N = n) \wedge H \sim Po(1).$$

Show that $E(B|(N = n) \wedge H) = n$ and therefore $E(B|H) = 13$. Obtain $var(B|H)$.

5. Prove that for any constants a, b, c, d, any random variables X, Y and information H,

$$I(aX + b, cY + d|H) = I(X, Y|H).$$

7
Continuous distributions

Random variables defined as counts of things naturally take integer values –
$0, 1, 2, \ldots$ – and the methods of Chapter 5 were implicitly developed for random
variables of this type. However, it is more usual in many problem areas for random variables to vary continuously. The lifetime of a light-bulb, for instance,
will not be an exact number of hours. The space between one possible value
and the next is not one hour, one minute, one second or even one millisecond.
All times are theoretically possible. This chapter deals with random variables
whose possible values are continuous.

Sections 7.1 to 7.3 are concerned with basic definitions. Continuous random
variables are introduced in Section 7.1 and problems of defining and measuring
probabilities are discussed. Section 7.2 presents a formal definition in terms of a
new function, the *distribution function* of a random variable, which is applicable
for both continuous and discontinuous random variables. Section 7.3 defines the
density function, which for continuous distributions is analogous to the probabil-
ities $P(X=x|H)$. Expectations are defined in Section 7.3, and Section 7.4 con-
siders transformations of continuous random variables. Measurement of distributions for continuous random variables is similar to the processes described in
Chapter 5. Some standard continuous distributions are defined in Section 7.5.

Sections 7.6 and 7.8 discuss joint distributions. In Section 7.6 we consider
two continuous random variables, and an extended example is analysed in Section 7.7. Section 7.8 deals with the case where one random variable is continuous and the other is discontinuous. Finally, Section 7.9 derives some important
distribution theory.

7.1 Continuous random variables

There are a large number of cows in Great Britain. I know that a popular breed
for dairy cows is Friesian, but what proportion of cows are Friesians? Let
$X=$ the proportion of dairy cows in Great Britain at this moment that are Friesians. Obviously, X lies between zero and one. If the number of dairy cows in
Great Britain is n then the possible values of X are $0, 1/n, 2/n, \ldots, {n-1}/n$ and 1.
The value of n is unknown, but is obviously large, so my knowledge allows X
to have practically all numbers in the interval $[0, 1]$ as its possible values. In
this section we discuss how I might define and measure my distribution for X.

Suppose that n is known, and for the moment we will make it absurdly small,
e.g. $n=7$. I could then consider my distribution for X by measuring my probabilities for its eight possible values, leading to the distribution shown in Figure
7.1. Now suppose that $n=14$. I might now give the distribution shown in Figure

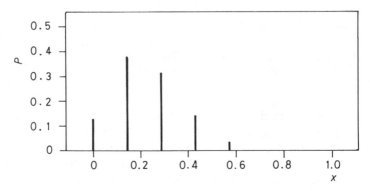

Figure 7.1. The Friesian cows example: $n=7$

7.2. Comparing the two distributions there are clear differences but also clear similarities. In Figure 7.2 there are nearly twice as many possible values. Since the probabilities in a distribution sum to one, the probabilities shown in Figure 7.1 are on average about twice as large as those in Figure 7.2. These are obvious and inescapable differences. However, both distributions show the same shape. Both are unimodal with a mode around $x=2/7$, and probabilities fall away at similar rates on either side of the mode.

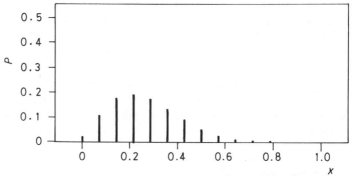

Figure 7.2. The Friesian cows example: $n=14$

If n were increased further, I would want to retain this shape. Since the average size of each probability would continue to fall as n increased, I might need a magnifying glass, or even a microscope, to see the shape, but the same shape should persist nevertheless. What happens if we let n grow indefinitely? As $n \to \infty$ there become infinitely many possible values and their average size tends to zero. Ultimately, all the probabilities will be zero and it will be impossible to maintain the shape, or even the condition that they sum to one. If we let the possible values of X be literally all numbers in $[0, 1]$ then the way in which we defined and measured probability distributions in Chapter 5 breaks down. With

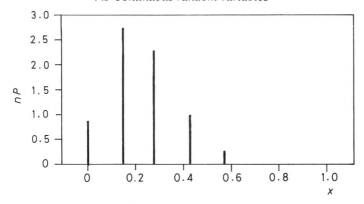

Figure 7.3. The Friesian cows example: $n=7$, scaled by n

so many possible values the probability that X will take any precise value is zero. Random variables whose possible values comprise all the numbers in some interval are called *continuous*, and we need to develop new ways of working with them.

$X =$ proportion of Friesians is not strictly continuous. Whilst n may be very large I could put an upper limit on it, say $n < 10^{10}$. I am sure that there are fewer than ten thousand million cows in Great Britain! Then the possible values of X are all the fractions with denominators less than 10^{10}. With this constraint, the probabilities will not degenerate completely to zero, and so we can retain the required shape. On the other hand, if we can manage to develop good ways of handling continuous random variables then, for the purpose of working with X it may be easier to treat it as continuous than to apply some arbitrary limit on n.

Our discussion of Figures 7.1 and 7.2 suggests how we might work with continuous random variables. We suggested that as n increased we might need a magnifying glass to see the shape. Let us redraw these diagrams, as Figures 7.3 and 7.4, to show $n P (X=x|H)$ instead of $P (X=x|H)$.

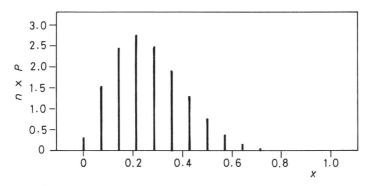

Figure 7.4. The Friesian cows example: $n=14$, scaled by n

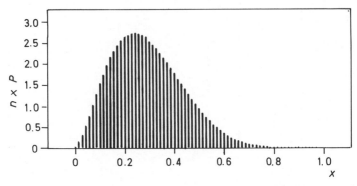

Figure 7.5. The Friesian cows example: $n=70$, scaled by n

Multiplying by n makes the average size of the probabilities stay about the same. Figures 7.3 and 7.4 now show more clearly how the shape persists, and we can now show the same phenomenon for much larger n, as in Figure 7.5, where $n=70$. Using a 'magnifying glass' brings out the shape, and will continue to do so as $n \rightarrow \infty$. As n increases there will be more and more lines filling out the shape until ultimately there will be a solid mass, bounded by a smooth curve. I could then measure my probabilities for X by stating the curve, for instance the curve shown in Figure 7.6.

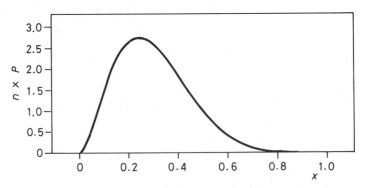

Figure 7.6. The Friesian cows example: large n, scaled

In Sections 7.2 and 7.3 we develop these ideas so that the distribution of a continuous random variable will indeed be represented by such a curve, known as its *density function*. Notice, however, that a careful approach is necessary because if we literally let n go to infinity then we are using an infinite amount of magnification to see the curve. The probabilities are all zero and the shape described by the density function cannot strictly be seen in the probabilities themselves. We must therefore be mathematically precise about what the density function represents.

7.2 Distribution functions

We cannot work with individual probabilities for continuous random variables because they are all strictly zero. Instead we consider the probability of a random variable lying in a given range. For $X =$ proportion of Friesians, it is meaningful to ask for my probability that X lies between 0.1 and 0.2, whatever the value of n. Whether there are a few, many or an infinite number of possible values in $[0.1, 0.2]$, I can give a non-zero probability for the proposition that $0.1 \le X \le 0.2$. For continuous X we cannot define its distribution by the set of probabilities $P(X = x | H)$, which are all zero, but we can use $P(X \le x | H)$, since these are genuine probabilities. Working in terms of $P(X \le x | H)$ allows random variables of all kinds to be handled in the same framework.

The distribution function

The distribution function of the random variable X given H is denoted by $D_{X|H}$ and is defined as

$$D_{X|H}(x) \equiv P(X \le x | H). \tag{7.1}$$

The distribution function is defined for all numbers x, not merely those x which are possible values of X.

Returning to the example of Friesian cows, consider the distribution function corresponding to the distribution shown in Figure 7.1, where $n = 7$. For all $x < 0$, $P(X \le x | H) = 0$, so (7.1) is zero at all these points. At $x = 0$, $P(X \le x | H) = P(X = 0 | H) > 0$, so $D_{X|H}(x)$ jumps at $x = 0$ to the same height as the first bar in Figure 7.1. It then stays at this value until $x = 1/7$, where it jumps up again. Its value now is $P(X \le 1/7 | H) = P(X = 0 | H) + P(X = 1/7 | H)$, the sum of the heights of the first two bars in Figure 7.1. It jumps again at $x = 2/7, 3/7, \ldots, 6/7$, and finally at $x = 1$. For all $x \ge 1$, the proposition '$X \le x$' is certain and so $D_{X|H}(x) = 1$. This is shown in Figure 7.7. Figures 7.8 and 7.9 show the distribution functions for $n = 14$ and $n = 70$. Again, as n increases the picture is developing into a smooth curve. The advantage of this series of pictures over Figures 7.3 to 7.6 is that the graphs represent actual probabilities, and continue to do so as n goes to infinity. What distinguishes a continuous random variable is that its distribution function is literally a smooth curve.

Random variables like those considered in Chapter 5, whose distributions may be defined through the set of probabilities $P(X = x | H)$, are called *discrete*. Their distribution functions look like Figures 7.7 and 7.8, where the function jumps up at each possible value and remains constant between possible values. This kind of function is called a *step function*.

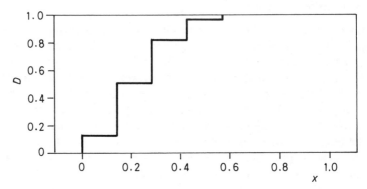

Figure 7.7. Friesian cows: $n=7$, distribution function

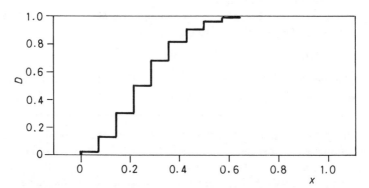

Figure 7.8. Friesian cows: $n=14$, distribution function

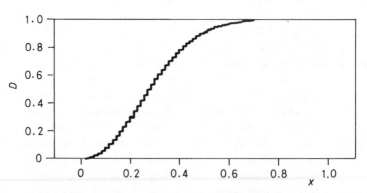

Figure 7.9. Friesian cows: $n=70$, distribution function

Discrete random variables

A random variable X is called discrete if its distribution function is a step function. Each point x at which the distribution function steps up is a possible value of X, and the height of the step in $D_{X|H}$ at x equals $P(X=x|H)$.

Notice that the discreteness of a random variable is not usually dependent on Your information. What does change with information is the heights, $P(X=x|H)$, of the steps. For a discrete random variable

$$D_{X|H}(a) = \sum_{x \le a} P(X=x|H),\tag{7.2}$$

with the sum being over all possible values x which are less than or equal to a. The distribution function and the set of probabilities $P(X=x|H)$ are equivalent ways of defining its distribution.

The distribution function of a continuous random variable can be considered as the limit of a series of step functions like Figures 7.7 and 7.8. It is a continuous, non-decreasing function with no steps.

Continuous random variables

A random variable is called continuous if its distribution function is a continuous function, i.e. has no steps.

{...........

Random variables of mixed type

It is possible for a distribution function to be neither a pure step function nor a purely continuous function, but to combine elements of both types. It may have some steps, at values which therefore have non-zero probabilities, but it may not be constant between steps. The fact that it rises smoothly between steps shows that there are possible values between steps but individually they have zero probabilities. We shall not encounter such distributions in this book, and will confine attention to random variables which are either purely discrete or purely continuous.

...........}

Exercises 7(a)

1. Calculate the distribution function of Bi(5, 0.3).

2. A continuous random variable S has distribution function $D_{S|H}(s)=0$ for $s \leq 0$, s^2 for $0 < s \leq 1$ and 1 for $s > 1$. Find $P(S \leq 0.3 | H)$ and $P(0.2 < S \leq 0.8 | H)$.

3. X is a continuous random variable taking values between 1 and 2. Consider the following functions:

$f_1(x) = (x-1)(2-x)$,
$f_2(x) = -\cos(\pi x/2)$,
$f_3(x) = e^x - 3$,

for $1 \leq x \leq 2$, and define $f_i(x) = 0$ for $x < 1$ and $f_i(x) = 1$ for $x > 2$ ($i = 1, 2, 3$). Which $f_i(x)$ could not be the distribution function of X?

7.3 Density functions

The distribution function defines and determines a distribution, but for discrete random variables it is more useful and more natural to work with the individual probabilities $P(X=x|H)$. To put this another way, we are less interested in the height of $D_{X|H}(x)$ than in the amount by which it jumps at x. Converting this into an analogous notion for a continuous distribution we could focus on the rate at which $D_{X|H}(x)$ is increasing at x. We therefore differentiate $D_{X|H}(x)$ with respect to x, and this derivative is known as the *density function*.

The density function
The density function of a continuous random variable X given H is denoted by $d_{X|H}$ and is defined as

$$d_{X|H}(x) \equiv \frac{\mathrm{d}}{\mathrm{d}x} D_{X|H}(x) . \tag{7.3}$$

The density function cannot be defined for a discrete random variable because we cannot differentiate its distribution function at the jump points. It is also possible for a continuous distribution function to have points at which it is not differentiable, but we shall ignore this detail and assume that $D_{X|H}(x)$ is differentiable at all x.

Since the distribution function is always non-decreasing, $d_{X|H}(x) \geq 0$ for all x. The significance of the density function is that it is precisely the function shown in Figure 7.6 for the Friesian cows example, the limit of the process represented by Figures 7.3 to 7.5. To see this, remember the definition of differentiation, whereby $d_{X|H}(x)$ is itself a limit,

$$d_{X|H}(x) = \frac{\mathrm{d}}{\mathrm{d}x} D_{X|H}(x) = \lim_{\delta x \to 0} \frac{D_{X|H}(x+\delta x) - D_{X|H}(x)}{\delta x} . \tag{7.4}$$

To relate (7.4) to the random variable X = proportion of Friesian cows, let

$$\delta x = -\frac{1}{n},$$

so that $n \to \infty$ is equivalent to $\delta x \to 0$. For any given value of n the possible values of X are $0, \frac{1}{n}, \ldots, 1$, and at any of these possible values

$$D_{X|H}(x - \frac{1}{n}) - D_{X|H}(x) = -P(X = x|H).$$

Substituting into (7.4) we find

$$d_{X|H}(x) = \lim_{n \to \infty} - \frac{P(X = x|H)}{-\frac{1}{n}} = \lim_{n \to \infty} n\, P(X = x|H).$$

It is precisely this limit that we considered in Section 7.1. We have now defined the density function in a more general and mathematically precise way, that is suitable for all continuous random variables.

The values of a distribution function are probabilities, but values of the density function are not. Strictly, the density at a point represents only the gradient of the distribution function at that point. However, we can obtain a more direct interpretation of the density function in terms of probabilities by reversing the definition (7.3). The inverse operation to differentiation is integration, and it follows immediately from (7.3) that

$$\int_a^b d_{X|H}(x)\, dx = D_{X|H}(b) - D_{X|H}(a) = P(a < X \leq b|H),$$

using (7.1).

Integrating the density function

For any continuous random variable X and information H, and any $a < b$,

$$\int_a^b d_{X|H}(x)\, dx = P(a < X \leq b|H). \tag{7.5}$$

In particular,

$$\int_{-\infty}^a d_{X|H}(x)\, dx = D_{X|H}(a), \tag{7.6}$$

$$\int_{-\infty}^\infty d_{X|H}(x)\, dx = 1. \tag{7.7}$$

In other words, the relationship between density and probability is that the area under the density function between any two points is a probability. The area under the whole density function is always one. These facts are illustrated in Figure 7.10.

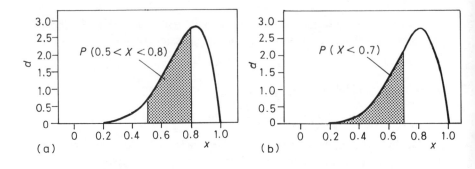

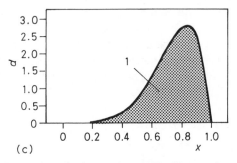

Figure 7.10. Examples of (a) equation (7.5), (b) equation (7.6), (c) equation (7.7)

{...........

Example

A random variable X has distribution function

$$D_{X|H}(x) = \begin{cases} e^x/2 & \text{if } x \leq 0, \\ 1 - e^x/2 & \text{if } x \geq 0. \end{cases}$$

This is shown in Figure 7.11(a). Its density function, shown in Figure 7.11(b), is

$$d_{X|H}(x) = \begin{cases} e^x/2 & \text{if } x \leq 0, \\ e^{-x}/2 & \text{if } x \geq 0 \end{cases}$$

$$= \exp(-|x|)/2. \tag{7.8}$$

(We use the fact that $de^x/dx = e^x$, a characteristic property of the exponential function.) Now consider the integral $\int_{-\infty}^{b} d_{X|H}(x)\,dx$. If $b \leq 0$ we find

$$\int_{-\infty}^{b} d_{X|H}(x)\,dx = \frac{1}{2}\int_{-\infty}^{b} e^x\,dx = \frac{1}{2}\left[e^x\right]_{-\infty}^{b} = e^b/2 = D_{X|H}(b)$$

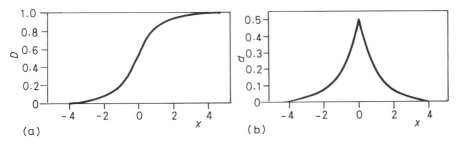

Figure 7.11. (a) Distribution function and (b) density function (Example)

since $\exp(-\infty) = 0$. For $b > 0$,

$$\int_{-\infty}^{b} d_{X|H}(x)\,dx = \frac{1}{2}\int_{-\infty}^{0} e^x\,dx + \frac{1}{2}\int_{0}^{b} e^{-x}\,dx = \frac{1}{2}\left[e^x\right]_{-\infty}^{0} + \frac{1}{2}\left[-e^{-x}\right]_{0}^{b}$$

$$= (1/2) + (-e^{-b} + 1)/2 = 1 - e^{-b}/2 = D_{X|H}(b)$$

again. Letting $b \to \infty$, $e^{-b} \to 0$ and the integral tends to 1, agreeing with (7.7).

............}

Because of these results, densities are like probabilities in their interpretation. A point with high density is, in a certain sense, more probable than one with low density. In the absolute sense, of course, both are so improbable as to have zero probability, but the probability that X lies 'near to' x is indicated by $d_{X|H}(x)$. X is more likely to be near a point with high density than one with low density. It should, however, be remembered that densities are only relative. Probabilities cannot exceed one, but densities are unrestricted. The constraint (7.7) is on the total area under the density function, therefore $d_{X|H}(x)$ can be arbitrarily large, provided that the range of x over which it is so large is correspondingly small.

Equations (7.5) to (7.7) also show how $d_{X|H}(x)$ for a continuous random variable plays a very similar role to that of $P(X=x|H)$ for a discrete random variable. Wherever we integrate $d_{X|H}(x)$ we can sum $P(X=x|H)$ for a discrete random variable to obtain the same probability. For instance, (7.6) and (7.2) express the distribution function as an integral or a sum respectively. Equation (7.7) is analogous to the fact that the probabilities in a discrete distribution must sum to one, and the discrete analogue of (7.5) is

$$\sum_{a < x \leq b} P(X=x|H) = P(a < X \leq b|H).$$

Many of the results in this chapter are translations of analogous results obtained in Chapters 5 and 6 for discrete random variables. We shall find that in almost every case the translation is very simple and consists of substituting $d_{X|H}(x)$ for $P(X=x|H)$ and replacing any summations by integrations.

7.4 Transformations and expectations

Equation (5.13) defines $E(X|H)$ for discrete X as a weighted average of possible values, with weights $P(X=x|H)$. For a continuous random variable we use $d_{X|H}(x)$ as the weight function. The total weight in both cases is one.

Expectation: continuous random variable

For a continuous random variable X, $E(X|H)$ is defined as

$$E(X|H) = \int_{-\infty}^{\infty} x \, d_{X|H}(x) \, dx \, . \tag{7.9}$$

As in the discrete case, the expectation may not exist, and wherever we use expectation there will be an implicit assumption of existence.

$var(X|H)$ is defined by equation (6.11). In order to determine whether equations (6.12) to (6.14) also apply in the continuous case we must first ask whether more basic results like (6.7) and (6.8) hold. This requires us to consider transformations of continuous random variables. Transformations are more complicated than in the discrete case. To appreciate the extra complexity, consider the simple linear transformation $Y = a + bX$, with $b > 0$. If X were discrete then for every possible value x of X corresponds a possible value $y = a + bx$ of Y, and $P(Y = a + bx|H) = P(X = x|H)$. Suppose, however, that X is continuous. Then Y is continuous and if X has as possible values all numbers in the range $[x_0, x_1]$ then the possible values of Y are all numbers in $[a + bx_0, a + bx_1]$. We cannot use $P(Y=y|H)$ or $P(X=x|H)$ since these are meaningless for continuous random variables. Consider instead

$$D_{Y|H}(y) = P(Y \le y|H) = P(a + bX \le y|H)$$

$$= P(X \le (y - a)/b|H)$$

$$= D_{X|H}((y-a)/b) \, , \tag{7.10}$$

using the fact that $b > 0$. Differentiating (7.10) with respect to y we must use the rule for differentiating a function of a function. We find

$$d_{Y|H}(y) = \frac{d}{dy} D_{X|H}((y-a)/b) = b^{-1} d_{X|H}((y-a)/b) \, . \tag{7.11}$$

The term b^{-1} in (7.11) arises as $d\{(y-a)/b\}/dy$. Because of this term, equation (7.11) is not a straight translation of the result for discrete random variables.

Now consider the transformation $Z = X^2$, which is not one-to-one. X^2 must be positive, so $D_{Z|H}(z) = 0$ for $z < 0$. Otherwise we find

$$D_{Z|H}(z) = P(X^2 \le z|H) = P(-z^{\frac{1}{2}} \le X \le z^{\frac{1}{2}}|H)$$

$$= D_{X|H}(z^{\frac{1}{2}}) - D_{X|H}(-z^{\frac{1}{2}}) \, .$$

$$\therefore d_{Z|H}(z) = (z^{-\frac{1}{2}}/2)\, d_{X|H}(z^{\frac{1}{2}}) - (-z^{-\frac{1}{2}}/2)\, d_{X|H}(-z^{\frac{1}{2}})$$

$$= z^{-\frac{1}{2}}\{ d_{X|H}(z^{\frac{1}{2}}) + d_{X|H}(-z^{\frac{1}{2}}) \}/2 . \tag{7.12}$$

Again, there is an extra term, $z^{-\frac{1}{2}}/2$, compared with the analogous result for discrete random variables.

Although it is possible to give a general result, expressing $d_{Y|H}$ in terms of $d_{X|H}$ when Y is an arbitrary transformation of X, both its statement and its proof are rather involved. In practice, it is often simpler and more instructive to proceed as in the above examples – first express $D_{Y|H}$ in terms of $D_{X|H}$ and then differentiate. However, the second example above shows that care must be taken in manipulating inequalities.

The following theorem is equally complicated to prove, and the proof is therefore omitted, but the theorem is simple to state.

Expectation of a continuous transformation
Let X be a continuous random variable, then for any information H and for any function g,

$$E\{g(X)|H\} = \int_{-\infty}^{\infty} g(x)\, d_{X|H}(x)\, dx . \tag{7.13}$$

{...........

Example
Suppose that X has the density (7.8). Its expectation is, from (7.9),

$$E(X|H) = \frac{1}{2} \int_{-\infty}^{\infty} x \exp(-|x|)\, dx$$

$$= \frac{1}{2} \int_{0}^{\infty} x\, e^{-x}\, dx + \frac{1}{2} \int_{-\infty}^{0} x\, e^{x}\, dx . \tag{7.14}$$

Both of these integrals can be done by parts. Thus

$$\int_{0}^{\infty} x\, e^{-x}\, dx = \left[x(-e^{-x}) \right]_{0}^{\infty} - \int_{0}^{\infty} (-e^{-x})\, dx$$

$$= 0 + \int_{0}^{\infty} e^{-x}\, dx = \left[-e^{-x} \right]_{0}^{\infty} = 1 , \tag{7.15}$$

$$\int_{-\infty}^{0} x\, e^{x}\, dx = \left[x\, e^{x} \right]_{-\infty}^{0} - \int_{-\infty}^{0} e^{x}\, dx = 0 - \left[e^{x} \right]_{-\infty}^{0} = -1 .$$

$$\therefore E(X|H) = 0 .$$

From Figure 7.11(b) it is not surprising that the mean of this distribution is zero. We now use the theorem (7.13) to find

$$E(X^2|H) = \frac{1}{2} \int_{-\infty}^{\infty} x^2 \exp(-|x|)\,dx .$$

Integrating by parts again gives the result. For instance

$$\int_0^{\infty} x^2 e^{-x}\,dx = \left[x^2(-e^{-x})\right]_0^{\infty} - \int_0^{\infty} (2x)(-e^{-x})\,dx$$

$$= 0 + 2\int_0^{\infty} x\,e^{-x}\,dx = 2 , \qquad (7.16)$$

using (7.15). Similarly, and by symmetry, $\int_{-\infty}^0 x^2 e^x\,dx = 2$, so that

$$E(X^2|H) = 2 . \qquad (7.17)$$

We could proceed to verify this result by formally letting $Y = X^2$, using (7.12) to obtain the density of Y given H and then applying (7.9) directly. Following this approach, but omitting the details, we find

$$E(X^2|H) = \frac{1}{2} \int_0^{\infty} y^{\frac{1}{2}} \exp(y^{-\frac{1}{2}})\,dy .$$

This integral can also be done by parts, using

$$\frac{d}{dy} \exp(y^{-\frac{1}{2}}) = -y^{-\frac{1}{2}} \exp(y^{-\frac{1}{2}}) .$$

After a second integration by parts we finally obtain (7.17).

............}

Equation (7.13) is a direct translation of (6.4). Therefore, although the density function of a continuous transformed random variable is more complicated than in the discrete case, its expectation follows an analogous and simple formula. It is now a simple matter to prove the theorem on the expectation of a linear function, equation (6.7), when X is continuous. In order to consider the expectations of sums and products of random variables we require further theory about joint distributions of continuous random variables, which is not presented until Sections 7.6 and 7.8. Anticipating that theory, however, it is easily established that equations (6.8) and (6.10) also apply unchanged when either X or Y is continuous, and when both are continuous. If we now define $var(X|H)$ by (6.11) then all the results (6.12) to (6.16) follow immediately. Therefore all of the expectation theory of Sections 6.3 and 6.4 is true regardless of whether we have discrete or continuous random variables.

Conditional distributions involving continuous random variables are also not defined until Sections 7.6 and 7.8, but it is easily shown that the expectation of an expectation theorem, equation (6.30) is true when either X or Y or both are continuous. The whole of Sections 6.5 to 6.7 are then found to be true for discrete and continuous random variables alike.

Exercises 7(b)

1. A random variable Z has density function

$$d_{Z|H}(z) = 2(1+z)^{-3}$$

for $z \geq 0$, and $d_{Z|H}(z) = 0$ for $z < 0$. Derive $D_{Z|H}(z)$, $P(Z > 1|H)$ and $E(Z|H)$.

2. Y is the proportion of people in a forthcoming election who will vote for the same candidate as You. You regard Y as essentially continuous and give it the density function

$$d_{Y|H}(y) = c \, y \, (1-y)$$

for $0 \leq y \leq 1$, and $d_{Y|H}(y)$ is zero otherwise. Determine the value of the constant c which makes this a proper density function. Obtain Your distribution function of Y and Your probability that Y lies between 0.4 and 0.6.

3. Prove that $E(aX+b|H) = a E(X|H) + b$, where X is a continuous random variable, and a and b are constants.

4. A random variable B has density function

$$d_{B|H}(b) = \begin{cases} \pi \cos(\pi b)/2 & \text{for } -1 \leq b \leq 1, \\ 0 & \text{otherwise}. \end{cases}$$

Verify that this is a proper density function, and find the density functions of $T \equiv 3B$, $S \equiv B^2$ and $E \equiv e^B$.

5. Why is it not possible to find the variance of the random variable Z in Exercise 1?

7.5 Standard continuous distributions

The process of measuring a distribution for a continuous random variable consists of measuring its density function. This cannot be done by separately measuring $d_{X|H}(x)$ at each point x, because there are infinitely many points to consider. Standard distributions are therefore essential. In this section we present some of the most important standard continuous distributions. In each case, as in Section 5.8, we give the shape, mean and variance. Measurement using standard distributions is the same as for discrete random variables. In particular, if the density is judged to be unimodal and not strongly skewed then the 'rules of thumb' for measuring mean and variance, given at the end of Section 5.8, apply also when X is continuous.

In the process of presenting the standard distributions, new ideas and new mathematical functions are discussed as they arise.

The continuous uniform family

The random variable Uc(a, b) is defined to have the uniform distribution on the interval [a, b], whose density function is

$$d_{Uc(a,b)}(x) = \begin{cases} (b-a)^{-1} & \text{, for } a \le x \le b \text{ ,} \\ 0 & \text{, otherwise .} \end{cases} \tag{7.18}$$

$$E(Uc(a,b)) = (b+a)/2 .$$

$$var(Uc(a,b)) = (b-a)^2/12 .$$

The distribution and density functions of a typical continuous uniform distribution are shown in Figure 7.12. Notice first that the distribution function is not differentiable at $x=a$ or $x=b$, because the gradient changes sharply at these points. The density function, which in Figure 7.12(b) seems to jump at these values, is not actually defined at $x=a$ or $x=b$. The problem is not important in practice because whatever we define the density to be at any isolated point, the integral (i.e. the area under the density function) between any two points is unchanged. In (7.18) we have arbitrarily defined the density function to be $(b-a)^{-1}$ at both points.

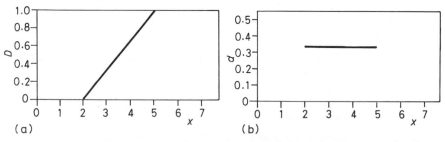

Figure 7.12. (a) Distribution function and (b) density function of Uc(2, 5)

The probability that Uc(a, b) lies in any range outside the interval [a, b] is zero, and we can therefore consider its possible values to consist only of those numbers between a and b. When evaluating integrals, such as in finding the expectation, any part of the range of integration lying outside [a, b] can be ignored. The contribution to the integral from outside this interval is always zero. For instance, if $a < c < b$ we have

$$D_{Uc(a,b)}(c) = \int_{-\infty}^{c} d_{Uc(a,b)}(x)dx$$

$$= \int_{-\infty}^{a} 0\,dx + \int_{a}^{c} (b-a)^{-1}dx \tag{7.19}$$

$$= \int_a^c (b-a)^{-1}dx = (c-a)/(b-a) ,$$

because the first term in (7.19) is zero. This fact is used below when deriving the mean and variance, to restrict the range of integration from $\int_{-\infty}^{\infty}$ to $\int_a^b$.

{...........

Proof

$$E\,(Uc(a,b)) = \int_a^b x\,(b-a)^{-1}dx = (b-a)^{-1}\left[x^2/2\right]_a^b$$

$$= \frac{1}{2}\,(b^2-a^2)/(b-a) = (b+a)/2 .$$

$$E\,(Uc(a,b)^2) = \int_a^b x^2\,(b-a)^{-1}dx = (b-a)^{-1}\left[x^3/3\right]_a^b$$

$$= \frac{1}{3}\,(b^3-a^3)/(b-a) = (b^2+ab+a^2)/3 ,$$

$$\therefore var\,(Uc(a,b)) = \frac{1}{12}\,\{\,4(b^2+ab+a^2) - 3(b+a)^2\,\}$$

$$= \frac{1}{12}\,(b^2-2ab+a^2) = (b-a)^2/12 .$$

...........}

{...........

Ignorance

Continuous uniform distributions are sometimes used to represent the situation where You know that X must lie between two bounds, a and b, but You otherwise have no knowledge to suggest that any value in this range is more probable than any other. The argument is perfectly valid when dealing with the discrete uniform family, but must be used with caution in the continuous case. For if You know only that X lies in $[a,b]$ (and suppose $a \geq 0$) then defining $Y \equiv X^2$, You know only that Y lies in $[a^2,b^2]$. If $X|H \sim Uc(a,b)$ it is not true that $Y|H \sim Uc(a^2,b^2)$. The properties of continuous transformations are more complex than with discrete random variables, and even one-to-one transformations of continuous uniform random variables are not generally uniform.

It is not enough to say loosely that You 'have no knowledge to suggest that any value in $[a,b]$ is more probable than any other'. This is meaningless for a continuous random variable. It is necessary to say that any subinterval $[c,d]$ has the same probability as any other subinterval of length $d-c$. Nonlinear transformations distort uniformity because they distort lengths of intervals.

...........}

The gamma family

The random variable $\text{Ga}(a, b)$ is defined to have the gamma distribution with parameters a and b, whose density is

$$d_{\text{Ga}(a,b)}(x) = \begin{cases} a^b x^{b-1} e^{-ax} / \Gamma(b) & , \text{ for } x \geq 0, \\ 0 & , \text{ for } x < 0. \end{cases} \tag{7.20}$$

The parameters a and b are both positive.

$$E(\text{Ga}(a, b)) = b/a . \tag{7.21}$$

$$var(\text{Ga}(a, b)) = b/a^2 . \tag{7.22}$$

The strange symbol Γ appearing in (7.20) denotes the *gamma function*. It is defined by

$$\Gamma(p) \equiv \int_0^\infty x^{p-1} e^{-x} dx .$$

If p is a positive integer then we can evaluate this integral by repeatedly integrating by parts, as in (7.15) and (7.16). Except for certain other cases, the integral cannot actually be evaluated exactly when p is not an integer. Like the mathematical constants π and e, it can only be computed numerically to a suitable number of decimal places. Therefore, like π and e, mathematicians represent it by a special symbol, $\Gamma(p)$.

The gamma function

The gamma function is denoted by Γ and is defined as

$$\Gamma(p) \equiv \int_0^\infty x^{p-1} e^{-x} dx$$

for $p > 0$. The following facts may be proved.

(a) $\Gamma(1) = 1$,   (b) $\Gamma(1/2) = \pi^{1/2}$,

(c) $\Gamma(p+1) = p\,\Gamma(p)$. $\tag{7.23}$

{..........

Proof

(a) is by direct integration. The proof of (b) is not simple, and we omit it. (c) is proved by integrating by parts, generalizing (7.15) and (7.16).

..........}

For integer n, we have

$$\Gamma(n+1) = n\ \Gamma(n) = n\,(n-1)\,\Gamma(n-1) = \cdots$$

$$= n\,(n-1)\,\cdots\,1\,\Gamma(1) = n!\ ,$$

therefore the gamma function can be considered as a generalization of the factorials.

For $b \le 1$ the gamma distributions are J-shaped; otherwise they are unimodal and skewed to the right. As b increases the degree of skewness decreases. The parameter a does not influence shape but simply scales the distribution. It is easy to show that if c is a positive constant then

$$c\ \mathrm{Ga}(a,b)\ \sim\ \mathrm{Ga}(a/c,b)\ . \tag{7.24}$$

Figure 7.13 shows the shapes of a selection of gamma density functions. Figure 7.13(b) and (d) demonstrate (7.24) with $c=10$, $a=1$ and $b=4$. The shapes are identical but the scale on the x axis in (d) is ten times that in (b).

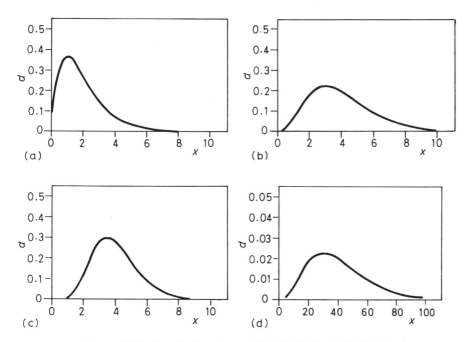

Figure 7.13. Density functions of (a) $\mathrm{Ga}(1,2)$, (b) $\mathrm{Ga}(1,4)$, (c) $\mathrm{Ga}(2,8)$, (d) $\mathrm{Ga}(0.1,4)$

{...........
Proof of (7.21) and (7.22)

$$E\left(\mathrm{Ga}(a,b)\right) = \int_0^\infty \{\, a^b\, x^b\, e^{-ax}/\Gamma(b)\,\}\, dx$$

$$= \{\,\Gamma(b+1)/a\,\Gamma(b)\,\} \int_0^\infty \{\, a^{b+1} x^b\, e^{-ax}/\Gamma(b+1)\,\}\, dx\ .$$

The integral is the integral of the Ga $(a, b+1)$ density over 0 to ∞ (or equivalently over $-\infty$ to ∞) and is therefore one. (Alternatively, change the variable from x to $y = ax$; the integral reduces to a gamma function.)

$$\therefore E\left(\mathrm{Ga}(a,b)\right) = \Gamma(b+1)/a\,\Gamma(b) = b/a\ ,$$

using (7.23). Similarly, $E\left(\mathrm{Ga}(a,b)^2\right) = \Gamma(b+2)/a^2\Gamma(b) = b\,(b+1)/a^2$,

$$\therefore var\left(\mathrm{Ga}(a,b)\right) = \{\, b\,(b+1)/a^2\,\} - b^2/a^2 = b/a^2\ .$$
..........}

The beta family
The random variable $\mathrm{Be}(p,q)$ is defined to have the beta distribution with parameters p and q, whose density function is

$$d_{\mathrm{Be}(p,q)}(x) = \begin{cases} x^{p-1}(1-x)^{q-1}/B(p,q)\ , & \text{if } 0 \le x \le 1, \\ 0 & \text{, otherwise}. \end{cases} \tag{7.25}$$

The parameters p and q are both positive.

$$E\left(\mathrm{Be}(p,q)\right) = p/(p+q)\ .$$

$$var\left(\mathrm{Be}(p,q)\right) = pq/\{\,(p+q)^2(p+q+1)\,\}\ .$$

The expression $B(p,q)$ in (7.25) denotes the beta function, which is also defined as an integral, and is related to the gamma function.

The beta function
The beta function with arguments p and q is denoted by $B(p,q)$ and defined as

$$B(p,q) = \int_0^1 x^{p-1}(1-x)^{q-1}\, dx\ , \tag{7.26}$$

for $p > 0$, $q > 0$. For any p, q,

$$B(p,q) = \Gamma(p)\,\Gamma(q)/\Gamma(p+q)\ . \tag{7.27}$$

The proof of (7.27) is omitted. Using (7.26) and (7.27) it is simple to verify the mean and variance of $\text{Be}(p,q)$.

Both parameters influence shape. If $p > 1$ and $q > 1$ the distribution is unimodal. It is symmetric if $p = q$, skewed to the right if $p > q$ and to the left if $q > p$. The amount of skewness increases with the discrepancy between p and q. If $p \leq 1$ and $q > 1$ the density is J-shaped, whereas if $p > 1$ and $q \leq 1$ it is the opposite of J-shaped, i.e. the density increases monotonically with x. If both $p \leq 1$ and $q \leq 1$ it becomes U-shaped, the opposite of unimodal, with the density first decreasing then increasing. As a special case, $\text{Be}(1,1) \sim \text{Uc}(0,1)$. The various shapes are displayed in Figure 7.14.

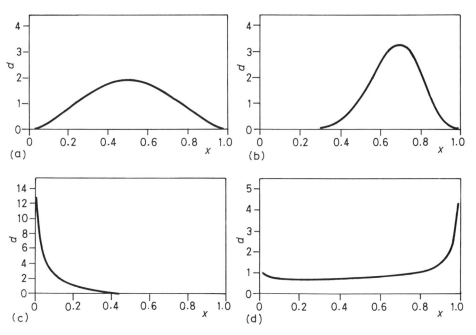

Figure 7.14. Density functions of (a) Be(3, 3), (b) Be(10, 5),
(c) Be(0.5, 6), (d) Be(0.8, 0.5)

Beta random variables take values in the interval $[0, 1]$, but this is easily modified by a linear transformation. The random variable $a + b\,\text{Be}(p,q)$ takes values in $[a, a+b]$, its shape is the same as that of $\text{Be}(p,q)$, and its mean and variance are easily found from equations (6.30) and (6.31).

The normal family

The random variable $N(m,v)$ is defined to have the normal distribution with mean m and variance v, whose density function is

$$d_{N(m,v)}(x) = (2\pi v)^{-\frac{1}{2}} \exp\{-(x-m)^2/(2v)\} \tag{7.28}$$

for all x. The parameter v is positive.

$$E(N(m,v)) = m .$$

$$var(N(m,v)) = v .$$

{...........

Proof

Consider

$$E(N(m,v)-m) = \int_{-\infty}^{\infty} (x-m)\, d_{N(m,v)}(x)\, dx$$

$$= \left[-(2\pi)^{-\frac{1}{2}} v^{\frac{1}{2}} \exp\{-(x-m)^2/(2v)\} \right]_{-\infty}^{\infty} = 0 .$$

$$\therefore E(N(m,v)) = E(N(m,v)-m) + m = m .$$

Next,

$$var(N(m,v)) = E(\{N(m,v)-m\}^2)$$

$$= \int_{-\infty}^{\infty} (x-m)^2\, d_{N(m,v)}(x)\, dx . \tag{7.29}$$

Consider first the half of the integration from m to ∞:

$$\int_{m}^{\infty} (x-m)^2 (2\pi v)^{-\frac{1}{2}} \exp\{-(x-m)^2/(2v)\}\, dx . \tag{7.30}$$

Make the change of variable

$$y = (x-m)^2/(2v) , \quad \therefore x = (2vy)^{\frac{1}{2}} + m , \quad \therefore \frac{dx}{dy} = v\,(2vy)^{-\frac{1}{2}} .$$

The integral (7.30) becomes

$$\int_{0}^{\infty} v\,(2vy)^{-\frac{1}{2}} (2vy) (2\pi v)^{-\frac{1}{2}} e^{-y}\, dy$$

$$= v\,\pi^{-\frac{1}{2}} \int_{0}^{\infty} y^{\frac{1}{2}} e^{-y}\, dy = v\,\pi^{-\frac{1}{2}} \Gamma(3/2)$$

$$= v\,\pi^{-\frac{1}{2}} \Gamma(1/2)/2 = v/2 .$$

We find similarly that the integral in (7.29) from $-\infty$ to m is also $v/2$, therefore $var\,(N(m,v))=v$.

.............}

Normal distributions are unimodal and symmetric. There is only one shape, and therefore the family is rather limited. Since linear transformations preserve shape, it is not surprising to find that any normal random variable may be transformed into any other by means of a linear transformation. In fact it is simple to show that

$$N(m,v) \sim m + v^{\frac{1}{2}}N(0,1). \tag{7.31}$$

The random variable $N(0,1)$ is known as the *standard normal* random variable, and all other normal random variables are derivable from it via (7.31). Figure 7.15 shows the density functions of the standard normal distribution and the $N(10,25)$ distribution. The latter is obtained from the former by shifting the center to $x=10$ and multiplying the x scale by $25^{\frac{1}{2}}=5$.

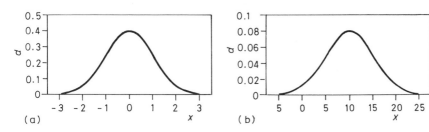

Figure 7.15. Density functions of (a) $N(0,1)$, (b) $N(10,25)$

The great strength of the normal family is its simplicity. Although the density function (7.28) may seem complicated, in mathematical terms it turns out to be very easy to work with. Consequently, there is a great wealth of distribution theory concerning normal random variables. Some will be presented in Sections 7.7 and 7.9.

Because of (7.31), the probability that $N(m,v)$ lies in any interval may be related to the probability that the standard normal random variable lies in some other interval. Thus

$$P(a \le N(m,v) \le b)$$
$$= P\left[v^{-\frac{1}{2}}(a-m) \le v^{-\frac{1}{2}}\{N(m,v)-m\} \le v^{-\frac{1}{2}}(b-m)\right]$$
$$= P\left[v^{-\frac{1}{2}}(a-m) \le N(0,1) \le v^{-\frac{1}{2}}(b-m)\right]$$
$$= \Phi(v^{-\frac{1}{2}}(b-m))-\Phi(v^{-\frac{1}{2}}(a-m)), \tag{7.32}$$

where Φ is the symbol generally used for the $N(0,1)$ distribution function, i.e. $\Phi(x) \equiv D_{N(0,1)}(x)$. Unfortunately this function, like the beta and gamma functions, is defined only as an integral,

$$\Phi(x) = \int_{-\infty}^{x} (2\pi)^{-\frac{1}{2}} \exp(-t^2/2)\,dt \, , \tag{7.33}$$

whose values must be obtained numerically. The same is true of the distribution functions of beta and gamma random variables. Even if b is an integer, so that $\Gamma(b)=(b-1)!$ is known exactly, the distribution function of $Ga(a,b)$ is not known and must be computed numerically. Because of (7.32), the distribution functions of all the normal random variables can be expressed in terms of Φ, the distribution function of $N(0,1)$. Values of $\Phi(x)$ for various x, and to various levels of numerical accuracy, may be obtained from several sources. First, they are given in any book of statistical tables. Until recently, such a book was a necessary tool of every statistician's trade. However, the development of computers has meant that most statisticians now have access to a more convenient source. Functions such as $\Phi(x)$ and $\Gamma(p)$ are available on most serious computer systems as pre-programmed routines. A few key-strokes suffices to obtain their values to a high degree of accuracy. For the reader who has neither of these sources available, a very short table of values is given below as Table 7.1.

x	0	0.5	1.0	1.5	2.0	2.5
$\Phi(x)$	0.5000	0.6915	0.8413	0.9332	0.9772	0.9938

Table 7.1. The standard normal distribution function

Note that, because of the symmetry of the normal density, $\Phi(-x)=1-\Phi(x)$, see Figure 7.16.

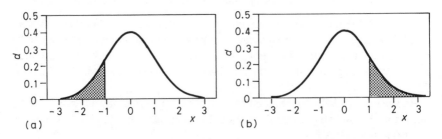

Figure 7.16. Symmetry of the normal distribution: (a) $\Phi(-1)$, (b) $1-\Phi(1)$

{...........

Two probabilities

Particular applications of (7.32) are

$$P(m-v^{\frac{1}{2}} \le N(m,v) \le m+v^{\frac{1}{2}}) = \Phi(1)-\Phi(-1) = 0.6826 \, ,$$

$$P(m-2v^{\frac{1}{2}} \le N(m,v) \le m+2v^{\frac{1}{2}}) = \Phi(2)-\Phi(-2) = 0.9544 \, ,$$

Notice that at the end of Section 5.8 the figures of 0.65 and 0.95 were offered as approximations to these probabilities (valid for any reasonably symmetric, unimodal distribution). Clearly, the approximations are good for normal distributions.

..........}

{..........
Example
Consider my distribution for X = the proportion of Friesian cows in Britain. I can measure my distribution as follows. First I estimate that X is about 0.3, so I assign $E_1(X|H) = 0.3$. I am not at all sure of my figure, and I am quite prepared to believe that X might be lower than 0.15 or higher than 0.45. This is consistent with a standard deviation of about 0.15 and, rounding slightly, this gives me $var_1(X|H) = 0.02$. Next, I judge that my distribution is unimodal. The natural choice of family for this distribution is the beta family, since these random variables are also defined on the interval $[0, 1]$. I must therefore give values to the parameters p and q consistent with my measured mean and variance. Therefore

$$p/(p+q) = 0.3, \quad pq/\{(p+q)^2(p+q+1)\} = 0.02$$

$$\therefore 0.3 \times 0.7 / (p+q+1) = 0.02$$

$$\therefore p+q = (0.21/0.02) - 1 = 9.5 .$$

$$\therefore p = 0.3 \times 9.5 = 2.85, \quad \therefore q = 6.65 .$$

My measured distribution is

$$X|H \ _1\sim \ Be(2.85, 6.65),$$

and is the distribution whose density is shown in Figure 7.6.

..........}

Exercises 7(c)

1. Prove that $E(Be(p, q)) = p/(p+q)$.

2. Let M be the amount of a mineral (in tonnes) which may be extracted from a new mine. The mining company estimates that M is around 500, but could possibly be as little as 100 or as large as 1500. Using the gamma family, obtain a suitable measured distribution.

3. A pilot measures his distribution for T = flying time (in hours) to his destination as N(4.4, 0.09). Obtain his measured probability for $T > 4.7$. What probability would he give to the flying time being between $4\frac{1}{4}$ and 5 hours?

4. Using standard distributions, measure your distribution for the following continuous random variables.

Of all wheat produced in 1985, W = the proportion grown in the U.S.A.

T = the melting point of tungsten.

M = the average height of adult men in Great Britain.

5. Let $X|H \sim \text{Be}(p,q)$ and suppose that $q > 2$. Find the density function, mean and variance given H of $Y \equiv X/(1-X)$. What problems are caused if $q \leq 2$?

6. Show that $\{N(0,1)\}^2 \sim \text{Ga}(0.5, 0.5)$.

7. Letting $Y \equiv \Phi\{N(0, 1)\}$, prove that $Y \sim \text{Uc}(0, 1)$.

8. A part of an electrical circuit has n identical components joined in series, so that the circuit fails if any one of the components fails. Let X_i be the length of time (in hours) for which component i works before failing, $i = 1, 2, \ldots, n$, and let S be the length of time that this part of the circuit works. Your belief about the X_is is that they are mutually independent with $X_i|H_1 \sim \text{Ga}(0.2, 1)$. Show that $P_1(X_i > x|H) = \exp(-0.2x)$. By considering $P_1(S > x|H)$ show that $S|H_1 \sim \text{Ga}(0.2n, 1)$.

7.6 Two continuous random variables

For any two random variables X and Y, we can define their joint distribution function given any information H.

Joint distribution function

The joint distribution function of random variables X and Y given H is denoted by $D_{X,Y|H}$ and defined as

$$D_{X,Y|H}(x,y) = P((X \leq x) \wedge (Y \leq y)|H). \qquad (7.34)$$

The joint distribution function can take a great variety of forms. The only constraint on it is that as either x or y increases $D_{X,Y|H}(x,y)$ cannot decrease. Certain forms, however, are extremely uncommon in practice. In this section and Section 7.8 we consider the most common forms.

If X and Y are both discrete then the possible values of (X,Y) are (x,y), where x is a possible value of X and y is a possible value of Y. Therefore $D_{X,Y|H}(x,y)$ will jump by an amount $P((X=x) \wedge (Y=y)|H)$ (which can, of course, be zero) at the point (x,y). Therefore, for discrete random variables the set of joint probabilities $P((X=x) \wedge (Y=y)|H)$ occur as heights of jumps in the distribution function. This is entirely analogous to the case of one random variable already discussed in Section 7.2. Formally, the relationship between $P((X=x) \wedge (Y=y)|H)$ and the joint distribution function is obtained as follows.

Suppose that (x,y) is a possible pair of (X,Y) values, and δx and δy are small, so that no other possible values occur between $x - \delta x$ and x or between $y - \delta y$ and y. Then

$$D_{X,Y|H}(x,y) - D_{X,Y|H}(x-\delta x,y) = P((X{=}x)\wedge(Y\le y)|H).\qquad(7.35)$$

Taking a corresponding difference in the y direction gives

$$\left[D_{X,Y|H}(x,y) - D_{X,Y|H}(x-\delta x,y)\right]$$
$$-\left[D_{X,Y|H}(x,y-\delta y) - D_{X,Y|H}(x-\delta x,y-\delta y)\right]$$
$$= P((X{=}x)\wedge(Y{=}y)|H).\qquad(7.36)$$

Between possible values of either x or y the distribution function is constant. Figure 7.17 shows a simple joint distribution function for two discrete random variables, as a three-dimensional perspective drawing.

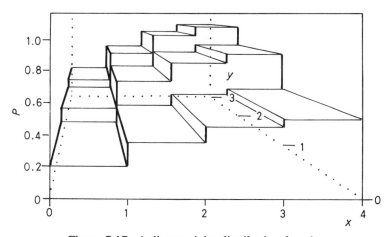

Figure 7.17. A discrete joint distribution function

It appears as a set of steps climbing in both x and y directions. Figure 7.18 shows the same distribution in a more abstract way. The number in each region is the value of $D_{X,Y|H}$ at every point within that region. Thus, $P((X\le x)\wedge(Y\le y)|H) = 0.56$ for all $1\le x < 2$ and $1\le y < 2$. The possible values of X are $0,1,2,3$ and the possible values of Y are $0,1,2$.

When X and Y are both continuous there are no jumps at all in their joint distribution function. The function becomes a smooth surface, climbing continuously as either x or y increases. Figure 7.19 shows a continuous joint distribution function as a perspective drawing. The lines criss-crossing the surface are plots of $D_{X,Y|H}(x,y)$ as one of its arguments varies while the other is held fixed. Figure 7.20 shows the same distribution function as a contour map. The lines here are contour lines. For instance, the line marked 0.5 joins all points (x,y)

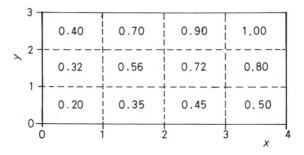

Figure 7.18. Diagram of a discrete joint distribution function

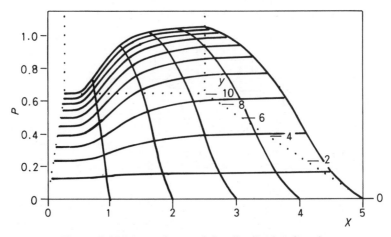

Figure 7.19. A continuous joint distribution function

for which $D_{X,Y|H}(x,y)=0.5$.

In order to define a joint density function we proceed analogously to (7.35) and (7.36). We need to differentiate rather than take differences, and we need to perform the operation in both x and y directions.

{.............

Partial derivatives.

Let f be a function of two variables, x and y. The partial derivative of f with respect to x is denoted by $\partial f / \partial x$ and defined as the limit

$$\frac{\partial f(x,y)}{\partial x} = \lim_{\delta x \to 0} \frac{f(x+\delta x, y)-f(x,y)}{\delta x}.$$

In other words, f is differentiated with respect to x alone, treating y as fixed. The partial derivative with respect to y is similarly defined.

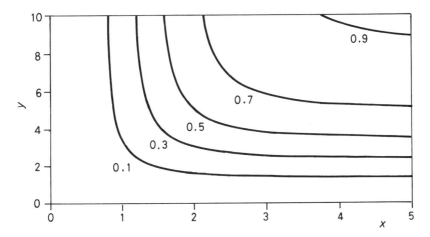

Figure 7.20. Contour map of a continuous joint distribution function

The partial derivative $\partial f / \partial x$ is a function of both x and y, and may also be differentiable with respect to either. Differentiating it partially with respect to y yields

$$\frac{\partial}{\partial y}\left(\frac{\partial}{\partial x}f(x,y)\right).$$

If this second differentiation is possible then (except in pathological cases) the same result may be achieved by differentiating in the opposite order. In such cases a simpler notation is used:

$$\frac{\partial^2}{\partial x \partial y}f(x,y) \equiv \frac{\partial}{\partial y}\left(\frac{\partial}{\partial x}f(x,y)\right) = \frac{\partial}{\partial x}\left(\frac{\partial}{\partial y}f(x,y)\right).$$

We shall only work with functions for which this result holds.

...........}

Joint density function
The joint density function of two continuous random variables X and Y, given H, is denoted by $d_{X,Y|H}(x,y)$ and defined as

$$d_{X,Y|H}(x,y) \equiv \frac{\partial^2}{\partial x \partial y} D_{X,Y|H}(x,y). \qquad (7.37)$$

Then

$$P((a < X \leq b) \wedge (c < Y \leq d)|H)$$

$$= \int_a^b \int_c^d d_{X,Y|H}(x,y)\, dx\, dy . \qquad (7.38)$$

{...........

Proof

Reversing the differentiation requires integration. Consider

$$\int_a^b d_{X,Y|H}(x,y)\,dx = \left[\partial D_{X,Y|H}(x,y)/\partial y\right]_a^b$$

$$= \frac{\partial}{\partial y}D_{X,Y|H}(b,y) - \frac{\partial}{\partial y}D_{X,Y|H}(a,y).$$

Now integrating with respect to y we have

$$\int_c^d \left[\int_a^b d_{X,Y|H}(x,y)\,dx\right]dy = \left[D_{X,Y|H}(b,y)\right]_c^d - \left[D_{X,Y|H}(a,y)\right]_c^d$$

$$= \left[D_{X,Y|H}(b,d) - D_{X,Y|H}(b,c)\right]$$

$$- \left[D_{X,Y|H}(a,d) - D_{X,Y|H}(a,c)\right]$$

$$= P((X \le b)\wedge(c < Y \le d)|H) - P((X \le a)\wedge(c < Y \le d)|H)$$

$$= P((a < X \le b)\wedge(c < Y \le d)|H).$$

The order in which the two integrals are performed is immaterial. (This is generally true of double integrals.) Therefore we use the more compact notation

$$\int_a^b \int_c^d d_{X,Y|H}(x,y)\,dx\,dy .$$

...........}

The two-dimensional integral (7.38) is calculating the volume under the surface defined by the density function and within the rectangle defined by $a < x \le b, c < y \le d$. The joint density function for continuous random variables acts like $P((X=x)\wedge(Y=y)|H)$ does for discrete random variables. High density at the point (x,y) is associated with relatively high probability that (X,Y) will take values in a small rectangle around (x,y).

We now wish to define marginal and conditional density functions. The marginal density function of X needs no new definition; it must be $d_{X|H}(x)$. Conditional densities are less obvious. In the discrete case the definition of the conditional distribution of Y given $(X=x)\wedge H$ is that it comprises simply the probabilities $P(Y=y|(X=x)\wedge H)$, which are all well-defined. In the continuous case it would be natural to start with a conditional distribution function of Y given $(X=x)\wedge H$. We might then expect it to satisfy

$$P(Y \le y|(X=x)\wedge H) = P((Y \le y)\wedge(X=x)|H)/P(X=x|H), \qquad (7.39)$$

but since X is continuous both the probabilities on the right-hand side are zero. Their ratio is meaningless without more detailed analysis. Therefore let us define $P(Y \leq y | (X=x) \wedge H)$ as the limit of $P(Y \leq y | (x < X \leq x + \delta x) \wedge H)$ as $\delta x \to 0$. Then the corresponding expression to (7.39) has both numerator and denominator non-zero, and the limit of their ratio can be properly defined as both tend to zero. Differentiating with respect to y will then provide a conditional density function.

Conditional probabilities and densities

If X is a continuous random variable then, for any proposition E and information H, define

$$P(E | (X=x) \wedge H) \equiv \lim_{\delta x \to 0} P(E | (x < X \leq x + \delta x) \wedge H). \qquad (7.40)$$

In particular, the conditional distribution functions of a random variable Y given $(X=x) \wedge H$ are defined by

$$D_{Y|X,H}(y,x) \equiv P(Y \leq y | (X=x) \wedge H). \qquad (7.41)$$

If Y is also a continuous random variable its conditional density functions given $(X=x) \wedge H$ are defined by

$$d_{Y|X,H}(y,x) \equiv \partial D_{Y|X,H}(y,x) / \partial y \qquad (7.42)$$

and satisfy

$$d_{X,Y|H}(x,y) = d_{X|H}(x) \, d_{Y|X,H}(y,x) \qquad (7.43)$$

for all x and y.

{ ....

Proof

Following the three definitions (7.40) to (7.42), equation (7.43) is proved as follows. Expanding (7.41) using (7.40) gives

$$D_{Y|X,H}(y,x) = \lim_{\delta x \to 0} \frac{P((Y \leq y) \wedge (x < X \leq x + \delta x) | H)}{P(x < X \leq x + \delta x | H)}$$

$$= \lim_{\delta x \to 0} \frac{D_{X,Y|H}(x+\delta x, y) - D_{X,Y|H}(x,y)}{D_{X|H}(x+\delta x) - D_{X|H}(x)}$$

$$= \lim_{\delta x \to 0} \frac{D_{X,Y|H}(x+\delta x, y) - D_{X,Y|H}(x,y)}{\delta x} \times$$

$$\lim_{\delta x \to 0} \frac{\delta x}{D_{X|H}(x+\delta x) - D_{X|H}(x)}$$

$$= \left(\frac{\partial}{\partial x} D_{X,Y|H}(x,y) \right) / d_{X|H}(x).$$

Differentiating with respect to y as in (7.42) and using (7.37) now gives

$$d_{Y|X,H}(y,x) = d_{X,Y|H}(x,y)/d_{X|H}(x)$$

and (7.43) is proved.

..........}

Although (7.40) is rather an abstract definition, it accords well with out notion of probability as a measure of degree of belief. For if X is continuous, You cannot really acquire the information that $X=x$. For example, if X is the weight in grams of a piece of fabric after a wear test You could observe a measurement of 44.6 grams, but You have not observed that $X=44.6$. The actual weight might be 44.61037725...., and the assertion that X is continuous implies that it cannot be determined with perfect accuracy. If the measuring instrument is correct to 0.1 gram then You have observed that $44.55<X<44.65$. Your belief about a proposition E might be slightly different if You used a more accurate weighing instrument and observed that $44.605<X<44.615$. But eventually further accuracy in measuring X would only change Your beliefs about E by tiny amounts. Ultimately we arrive at the limit (7.40). When we observe a continuous random variable in practice then, unless the observation is unusually inaccurate, it is safe to regard a measurement of $P(E|(x<X<x+\delta x)\wedge H)$ as more simply a measurement of $P(E|(X=x)\wedge H)$, ignoring the observation error.

Equation (7.43) is now a direct translation of the discrete form (5.3). We can therefore also define independence and develop extending the argument and Bayes' theorem as direct translations of discrete versions.

Independence, extending the argument and Bayes' theorem

Let X and Y be continuous random variables. They are said to be independent given H if

$$d_{X,Y|H}(x,y) = d_{X|H}(x)\,d_{Y|H}(y) . \tag{7.44}$$

In general, $d_{X|H}(x)$ may be elaborated by extending the argument to include Y, the formula being

$$d_{X|H}(x) = \int_{-\infty}^{\infty} d_{Y|H}(y)\,d_{X|Y,H}(x,y)\,dy . \tag{7.45}$$

The continuous version of Bayes' theorem is

$$d_{Y|X,H}(y,x) = \frac{d_{Y|H}(y)\,d_{X|Y,H}(x,y)}{d_{X|H}(x)} \tag{7.46}$$

$$= \frac{d_{Y|H}(y)\,d_{X|Y,H}(x,y)}{\int_{-\infty}^{\infty} d_{Y|H}(z)\,d_{X|Y,H}(x,z)\,dz} . \tag{7.47}$$

{...........}

Proof

To prove (7.45), the right hand side is

$$\int_{-\infty}^{\infty} d_{X,Y|H}(x,y)\,dy = \int_{-\infty}^{\infty} d_{X|H}(x)\,d_{Y|X,H}(y,x)\,dy$$

$$= d_{X|H}(x)\int_{-\infty}^{\infty} d_{Y|X,H}(y,x)\,dy = d_{X|H}(x),$$

since the integral from $-\infty$ to ∞ of any density function is one – equation (7.7). It is worth noting that a marginal density is obtained by integrating the joint density with respect to the unwanted variable, over the whole range $(-\infty, \infty)$. Bayes' theorem (7.46) is an immediate consequence of the asymmetry of (7.43), and substituting (7.45) gives (7.47).

{...........}

We conclude this section with a brief discussion of expectations. If $Z = g(X,Y)$ is a function of both X and Y, then $E(Z|H)$ may be obtained by

$$E(g(X,Y)|H) = \int_{-\infty}^{\infty}\int_{-\infty}^{\infty} g(x,y)\,d_{X,Y|H}(x,y)\,dx\,dy\;. \tag{7.48}$$

Notice that if g is a function only of one variable, say $g(X)$, (7.48) reduces to

$$E(g(X)|H) = \int_{-\infty}^{\infty}\int_{-\infty}^{\infty} g(x)\,d_{X|H}(x)\,d_{Y|X,H}(y,x)\,dx\,dy$$

$$= \int_{-\infty}^{\infty} g(x)\,d_{X|H}(x)\left[\int_{-\infty}^{\infty} d_{Y|X,H}(y,x)\,dy\right]dx$$

$$= \int_{-\infty}^{\infty} g(x)\,d_{X|H}(x)\,dx\;,$$

and is seen to be consistent with (7.13). Equation (6.8) for the expectation of a sum follows immediately. Equation (6.10) for the expectation of a product of independent random variables is also easily proved. Conditional expectations are defined as in

$$E(g(X)|y,H) = \int_{-\infty}^{\infty} g(x)\,d_{X|Y,H}(x,y)\,dx$$

and therefore the regression function of X on Y given H is $h(y) \equiv E(X|y,H)$. If we define $E(X|Y,H)$ to be $h(Y)$ then equation (6.30) is easily proved.

The following section consists of an extended illustration of the ideas and results of this section. It also introduces, in a simple form, the important concepts of statistical modelling and inference, which will be studied in Chapter 9.

7.7 Example: heat transfer

In a typical nuclear reactor, such as the British magnox reactors, heat generated within the nuclear core is collected by blowing a gas over the fuel elements. The heated gas is then used to drive turbines, and thereby generates electricity. It is important to achieve efficient heat transfer from fuel elements to gas. If at any point on any element the heat transfer is too low then there is a danger that the element will overheat and rupture, with possibly explosive consequences.

Heat transfer rates are difficult to predict theoretically. They depend intimately on the characteristics of the gas flow across the surface of the element. This will be different at different parts of the core, and is influenced by the gas flow rate, the geometry of the elements, how they are assembled, and so on. Therefore, heat transfer is explored experimentally, in laboratory conditions (with nuclear fuel replaced by electric heaters!). In practice, heat transfer will be measured at a multitude of points, under a variety of conditions, with a view to predicting average and minimum rates in a real core. We consider in this section a greatly simplified example, where there is only one set of conditions and only one point of interest. An experiment is performed and the heat transfer coefficient at this point is measured as 8.4. The physicist does not now believe that if these conditions were duplicated in a real reactor the heat transfer coefficient observed at this point would actually be 8.4. There are numerous reasons why it might be a different figure, such as experimental error, inherent variability in operating conditions (e.g. the vagaries of turbulent gas flow), and the difference between a laboratory experiment and reality. So, allowing for all these factors, what might the physicist believe?

Let C be the average heat transfer coefficient at this point over a suitable period of time in a real reactor. Let the physicist's information be H. She requires to measure her distribution for C given H. Now H includes the information that $M = 8.4$, where M denotes experimental measurement. (Actually M is 'close' to 8.4, allowing for measurement error.) So let $H = H_0 \wedge (M = 8.4)$.

Before performing the experiment, the physicist would have expected C to be about 9. The experiment suggests that it might be as low as 8.4. How should the two sources of information be balanced? The appropriate elaboration is Bayes' theorem. Her prior estimate of 9 corresponds to a measurement.

$$E_1(C|H) = 9 .$$

She now considers how far from 9 she would have expected C to lie (before doing the experiment). She considers that C could well be as low as 8 or as high as 10, but is highly improbable to be lower than 7 or higher than 11. This suggests that the range 8 to 10 comprises the mean plus and minus one standard deviation (probability about 0.65), and that 7 to 11 is the mean plus and minus two standard deviations (probability about 0.95). Hence,

$$var_1(C|H_0) = 1^2 = 1 .$$

She regards a symmetric, unimodal density as appropriate, and therefore

$$C|H_0 \ _1 \sim N(9,1) . \tag{7.49}$$

She now considers the adequacy of the experiment. If the true value of C were c she would expect M to be within plus or minus 0.5 of c. Again her distribution is selected to be unimodal and symmetric. These considerations lead her to

$$M|(C=c) \wedge H_0 \ _1 \sim N(c,0.25) . \tag{7.50}$$

Equations (6.49) and (6.50) together imply her joint density function for C and M. Thus,

$$
\begin{aligned}
d^1_{C,M|H_0}(c,m) &= d^1_{C|H_0}(c)\, d^1_{M|C,H_0}(m,c) \\
&= \left[(2\pi)^{-\frac12}\exp\{-(c-9)^2/2\} \right]\left[(\pi/2)^{-\frac12}\exp\{-2(m-c)^2\} \right] \\
&= \pi^{-1}\exp\left[-\{(c-9)^2+4(m-c)^2\}/2 \right] .
\end{aligned}
\tag{7.51}
$$

Since the subscripts on the letter d are important to indicate the various density functions, we use a superscript to show that they are measured.

We require the conditional distribution of C given $H=(M=8.4)\wedge H_0$. The natural elaboration using (7.49) and (7.50) is Bayes' theorem. The numerator of Bayes' theorem will be (7.51), and its denominator will be the (marginal) density of M given H_0. We therefore need to integrate (7.51) with respect to c. First notice that

$$(c-9)^2+4(m-c)^2 = 5c^2-18c-8mc+81+4m^2$$

$$= 5\left(c - \frac{9+4m}{5}\right)^2+q ,$$

where

$$q = 81+4m^2-(9+4m)^2/5 = 4(m-9)^2/5 .$$

Substituting into (7.51), the integral becomes

$$
\begin{aligned}
d^1_{M|H_0}(m) &= \int_{-\infty}^{\infty} d^1_{C,M|H_0}(c,m)\,dc \\
&= \pi^{-1}\exp\{-2(m-9)^2/5\} \times \\
&\qquad \int_{-\infty}^{\infty} \exp\{-5(c-(9+4m)/5)^2/2\}\,dc .
\end{aligned}
\tag{7.52}
$$

The integral in (7.52) is now the integral of the $N((9+4m)/5, 1/5)$ density function, apart from its initial constant. We can therefore do the integral immediately, obtaining

$$d^1_{M|H_0}(m) = \pi^{-1}\exp\{-2(m-9)^2/5\}\,(10\pi)^{\frac12} \tag{7.53}$$

$$\therefore M|H_0 \quad _1\sim \quad N(9, {}^5/4) \, . \tag{7.54}$$

This measurement is an example of extending the argument. Dividing (7.51) by (7.53) gives

$$d^1_{C|M,H}(c,m) = (10\pi)^{-\frac{1}{2}} \exp\{-5(c-(9+4m)/5)^2/2\} \, ,$$

$$\therefore C|(M=m)\wedge H \quad _1\sim \quad N((9+4m)/5, {}^1/5) \, . \tag{7.55}$$

Inserting $m=8.4$, the result (7.55) gives the physicist's required distribution

$$C|H \quad _1\sim \quad N(8.52, 0.2) \, . \tag{7.56}$$

Formulae (7.54) and (7.55) are special cases of a general theorem which is proved in Section 7.9.

Let us look more closely at what is happening in this example. The physicist starts out with the beliefs about C expressed in the distribution (7.49) and shown in Figure 7.21. She observes $M = 8.4$ and updates her belief about C by Bayes' theorem. As always, Bayes' theorem operates by multiplying the prior distribution by the likelihood function. In this case, the likelihood function is $d_{M|C,H}(m,c)$ but regarded as a function of c, with $m=8.4$. This function is also drawn in Figure 7.21. Bayes theorem takes the product, (7.51), then rescales it by the denominator, (7.53), so that it integrates to one. The resulting posterior distribution is (7.56), whose density is also shown in Figure 7.21.

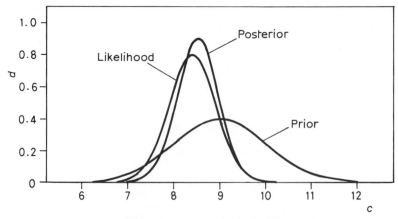

Figure 7.21. Prior, likelihood and posterior: heat transfer example

We can see in Figure 7.21 the two sources of information – prior and likelihood. The likelihood is tighter and therefore more informative than the prior. Consequently the posterior is centred on a value, 8.52, which is closer to the observation of 8.4 than to the prior mean of 9.

Equation (7.55) shows that this happens regardless of the observed value $M=m$. The posterior mean is $(9+4m)/5$, which is a weighted average of the prior mean 9 and the observation m. The observation gets weight $4/5$, while the prior weight is only $1/5$. The posterior mean will be closer to the data because the data information is stronger than the prior information in this case. The posterior variance is 0.2, smaller than the prior variance of 1 or the observation error variance of 0.25. Combining the two sources of information is reflected in a posterior belief which is stronger than either source separately.

Notice also that both regression functions are linear.

$$E_1(M|(C=c)\wedge H_0) = c .$$

$$E_1(C|(M=m)\wedge H_0) = (9/5) + (4/5)m .$$

Therefore the theory of Section 6.7 applies. We can deduce from (6.38) that

$$\frac{cov_1(M,C|H_0)}{var_1(C|H_0)} = 1 , \qquad \frac{cov_1(M,C|H_0)}{var_1(M|H_0)} = \frac{4}{5} .$$

Since $var_1(C|H_0)=1$, we have

$$cov_1(M,C|H_0) = 1 , \qquad var_1(M|H_0) = 5/4 ,$$

the latter agreeing with (7.54). Therefore

$$corr_1(M,C|H_0) = 2/5^{\frac{1}{2}} = 0.894 ,$$

a high correlation coefficient. From (6.39) its square is

$$0.8 = I_1(M,C|H_0) = var_1(C|H_0) - E_1\{var(C|M\wedge H_0)\}$$

$$\therefore E_1\{var(C|M\wedge H_0)\} = 0.2 .$$

This is certainly true of the present example. Indeed, (7.55) says that $var_1(C|(M=m)\wedge H_0)=0.2$ for all values of m. This feature, of the expected variance reduction given by the equation (6.39) being achieved for all possible values of the conditioning variable, is a characteristic of the kind of normal distribution structure shown in this example.

Exercises 7(d)

1. The joint distribution function of two random variables K and M is defined as follows.

$$D_{K,M|H}(k,m) = \begin{cases} km(k+m)/2 , & \text{if } 0\leq k \leq 1 \text{ and } 0\leq m \leq 1 ; \\ 0 , & \text{if } k<0 \quad \text{or } m<0 ; \\ 1 , & \text{if } k>1 \quad \text{and } m>1 ; \\ k(k+1)/2 , & \text{if } 0\leq k \leq 1 \text{ and } m>1 ; \\ m(m+1)/2 , & \text{if } k>1 \quad \text{and } 0\leq m \leq 1 . \end{cases}$$

Show that their joint density function is $d_{K,M|H}(k,m)=k+m$ if $0\leq k \leq 1$ and $0\leq m \leq 1$, and is otherwise zero. Obtain the marginal distribution functions

$D_{K|H}(k)$ and $D_{M|H}(m)$, and corresponding density functions. Using (7.43) derive the conditional density function of K given $H \wedge (M = 0.5)$.

2. A certain kind of impurity arises naturally when manufacturing metal pipes, and helps to inhibit corrosion. Let A be the amount of impurity in a specimen of the metal and let C be the amount of corrosion over a period of 24 hours in an autoclave, under conditions of high temperature and humidity. (Both are measured in suitable units.) The metallurgist's beliefs about A and C are expressed as follows. Given $A = a$, she expects that C will be about $100/a$ with standard deviation $10/a$, corresponding to $C | H \wedge (A = a) \mathbin{\sim_1} \mathrm{Ga}(a, 100)$. Her measured distribution for A is $\mathrm{Ga}(1, 10)$. Derive her marginal distribution for C. The experiment is performed and C is found to be 17. Obtain her conditional distribution for A given $H \wedge (C = 17)$.

3. Let X and Y be continuous random variables. The expectation of a function $g(X, Y)$ is defined to be

$$E\{g(X,Y)|H\} = \int_{-\infty}^{\infty} \int_{-\infty}^{\infty} g(x,y)\, d_{X,Y|H}(x,y)\, dx\, dy .$$

Prove that (6.8) holds, also (6.10) if X and Y are independent given H.

Suppose that g is a function only of X, i.e. $g(x, y) = h(x)$ for all y. Prove that the above definition is consistent with (7.9) by showing that in this case

$$E\{g(X,Y)|H\} = \int_{-\infty}^{\infty} h(x)\, d_{X|H}(x)\, dx .$$

7.8 Random variables of mixed type

Suppose now that X is continuous but Y is discrete. Their joint distribution function is still defined as in (7.34). If we hold x fixed and let y vary, $D_{X,Y|H}(x,y)$ is a step function, jumping up at possible values of y. If we hold y fixed it is a continuous, smoothly-increasing function of x. The amounts by which it steps up in the y direction are $P((X \leq x) \wedge (Y = y)|H)$. The equivalent of a joint density function is obtained if we now differentiate with respect to x. We will call the result the *joint semi-density* of X and Y given H.

Joint semi-density function

If X is a continuous random variable and Y is discrete, their joint semi-density function given H is denoted by $d_{X,(Y=y)|H}(x)$ and defined as

$$d_{X,(Y=y)|H}(x) = \frac{\partial}{\partial x} P((X \leq x) \wedge (Y = y)|H) . \qquad (7.57)$$

It follows from (7.57) that

$$P((a < X \le b) \wedge (Y=y) | H) = \int_a^b d_{X,(Y=y)|H}(x) \, dx \ . \tag{7.58}$$

Marginal and conditional distributions require no new concepts. Given $(Y=y) \wedge H$, X has a density function $d_{X|(Y=y) \wedge H}(x)$ defined in the usual way. The conditional distribution of Y given $(X=x) \wedge H$ contains the probabilities $P(Y=y | (X=x) \wedge H)$ defined by (7.40). Multiplying by the appropriate marginal density or probability gives two formulae for the joint semi-density function.

Independence for random variables of mixed type
If X is continuous and Y is discrete then

$$d_{X,(Y=y)|H}(x) = d_{X|H}(x) P(Y=y | (X=x) \wedge H) \tag{7.59}$$

$$= P(Y=y | H) d_{X|(Y=y) \wedge H}(x) \ . \tag{7.60}$$

X and Y are said to be independent given H if

$$d_{X,(Y=y)|H}(x) = d_{X|H}(x) P(Y=y | H) \ . \tag{7.61}$$

{...........

Proof

(i) $d_{X|H}(x) P(Y=y | (X=x) \wedge H)$

$$= \lim_{\delta x \to \infty} \frac{D_{X|H}(x+\delta x) - D_{X|H}(x)}{\delta x} \times$$

$$\lim_{\delta x \to \infty} \frac{P((X \le x+\delta x) \wedge (Y=y) | H) - P((X \le x) \wedge (Y=y) | H)}{D_{X|H}(x+\delta x) - D_{X|H}(x)}$$

$$= \lim_{\delta x \to \infty} \frac{P((X \le x+\delta x) \wedge (Y=y) | H) - P((X \le x) \wedge (Y=y) | H)}{\delta x}$$

$$= d_{X,(Y=y)|H}(x) \ .$$

(ii) $P(Y=y | H) d_{X|(Y=y) \wedge H}(x)$

$$= P(Y=y | H) \times$$

$$\lim_{\delta x \to \infty} \frac{P((X \le x+\delta x) | (Y=y) \wedge H) - P((X \le x) | (Y=y) \wedge H)}{\delta x}$$

$$= \lim_{\delta x \to \infty} \frac{P((X \le x+\delta x) \wedge (Y=y) | H) - P((X \le x) \wedge (Y=y) | H)}{\delta x}$$

$$= d_{X,(Y=y)|H}(x) \ .$$

...........}

Because of the asymmetry between X and Y, the 'multiplication law' takes two different forms, (7.59) and (7.60). The definition of independence, (7.61), is given in its symmetric form, but implies two different versions of the statement that learning one variable does not change your beliefs about the other. Specifically, X and Y are independent given H if either $d_{X|(Y=y)\wedge H}(x) = d_{X|H}(x)$ for every possible y, or $P(Y=y|(X=x)\wedge H) = P(Y=y|H)$ for every x. Similarly, extending the argument and Bayes' theorem each have two different forms.

Extending the argument: variables of mixed type

If X is continuous and Y is discrete then

$$P(Y=y|H) = \int_{-\infty}^{\infty} P(Y=y|(X=x)\wedge H)\, d_{X|H}(x)\, dx ,\qquad (7.62)$$

$$d_{X|H}(x) = \sum_y d_{X|(Y=y)\wedge H}(x)\, P(Y=y|H) .\qquad (7.63)$$

with the latter sum being over all possible values y of Y.

The proofs of (7.62) and (7.63) are straightforward. Essentially, we are regarding the joint semi-density as a two-way table and finding a margin of the table by summing in the appropriate direction. Except that in one direction we need to translate the sum into an integral.

Bayes' theorem: random variables of mixed type

If X is continuous and Y is discrete then

$$P(Y=y|(X=x)\wedge H)$$

$$= \frac{P(Y=y|H)\, d_{X|(Y=y)\wedge H}(x)}{\sum_{y'} P(Y=y'|H)\, d_{X|(Y=y')\wedge H}(x)} ,\qquad (7.64)$$

$$d_{X|(Y=y)\wedge H}(x)$$

$$= \frac{d_{X|H}(x)\, P(Y=y|(X=x)\wedge H)}{\int_{-\infty}^{\infty} d_{X|H}(z)\, P(Y=y|(X=z)\wedge H)\, dz} .\qquad (7.65)$$

Equations (7.64) and (7.65) follow immediately from (7.60) to (7.63). Finally, expectations are defined in the obvious way:

$$E(g(X,Y)|H) = \sum_y \int_{-\infty}^{\infty} g(x,y)\, d_{X,(Y=y)|H}(x)\, dx .\qquad (7.66)$$

The various theorems of Sections 6.3 to 6.7 are again easily proved to hold.

In Chapter 8 we shall make quite extensive use of the formulae of this section.

Exercises 7(e)

1. Consider the electrical circuit problem of Exercise 7(c)8 and suppose that the number of components is a random variable N. You believe that N is either 1, 2 or 3, and You assign these three values equal probabilities given H. Obtain the joint semi-density function of S and N given H, using (7.60). Derive the marginal density function of S. Show that

$$P_1(N=n\,|\,H \wedge (S=s)) = n\,\exp(-0.2ns)\Big/ \sum_{m=1}^{3} m\,\exp(-0.2ms)\,,$$

$(n=1,2,3)$. Calculate Your distribution for N given $H \wedge (S=5)$.

2. Let $X\,|\,H \sim \mathrm{Be}(a,b)$ and $Y\,|\,H \wedge (X=x) \sim \mathrm{Ge}(x)$. Write down the joint semi-density function of X and Y given H, and derive the (marginal) distribution of Y given H. Prove that

$$X\,|\,H \wedge (Y=y) \sim \mathrm{Be}(a+1, b+y)\,.$$

7.9 Continuous distribution theory

We noted earlier that the binomial distributions could be obtained as limits of hypergeometric distributions, by imagining deals from large collections (Section 4.11). We also noted that the Poisson distributions arose as limits of binomial distributions, in the context of events occurring randomly in time. Several other theorems exist of this type, in many of which the limiting distributions are the normal family. We present some of these later, but first we introduce the simplifying notion of standardization.

Standardization
Let X be a random variable with $E(X\,|\,H)=m$ and $\mathrm{var}(X\,|\,H)=v$. Then the standardized form of X is the linear transformation

$$Y = v^{-\frac{1}{2}}(X-m)\,, \tag{7.67}$$

which has $E(Y\,|\,H)=0$ and $\mathrm{var}(Y\,|\,H)=1$.

The standardizing transformation (7.67) can be applied to any random variable. We have seen in (7.31) that the standardized form of a N(m, v) random variable is the standard normal, N(0, 1), random variable.

Standardizing discrete random variables alters their possible values. For instance, the standardized form of Bi(n, p) is,

$$Y = (\text{Bi}(n,p) - np)/\{np(1-p)\}^{\frac{1}{2}},$$

whose possible values start at $-\{np/(1-p)\}^{\frac{1}{2}}$, then increase in steps of $\{np(1-p)\}^{-\frac{1}{2}}$ and stop at $\{n(1-p)/p\}^{\frac{1}{2}}$. Consider what happens to this random variable as $n \to \infty$. The upper and lower limits of the possible values tend to $\pm\infty$ and the steps between possible values tend to zero. Therefore as n increases, Y becomes more and more like a continuous random variable, taking all values from $-\infty$ to ∞. It can be shown that its distribution tends to that of N(0, 1). The same thing happens as we take limits of the standardized forms of various other random variables. We summarize these theorems below, without proof.

Normal limit theorems
The standardized forms of the following random variables have distributions which tend to that of N(0, 1) under the limits stated.

(1) Bi(n, p), as $n \to \infty$,
(2) Po(m), as $m \to \infty$,
(3) NB(n, p), as $n \to \infty$,
(4) Ga(a, b), as $b \to \infty$,
(5) Be(p, kp), as $p \to \infty$ for any $k > 0$.

All of these limit theorems are direct corollaries of a single theorem, the *central limit theorem*, which states in very general terms that the limiting distribution of the standardized form of a sum of random variable tends to that of $N(0,1)$ as the number of random variables in the sum tends to infinity. It is partly because of this theorem that the normal family plays a very important role in distribution theory and statistical analysis.

In practical terms, the normal limit theorems are a great aid to computing probabilities for many distributions. For instance suppose that $X|H \sim \text{Bi}(1000, 0.5)$ and we require $P(X < 520|H)$. Since the 'n' parameter is large ($n=1000$) this binomial distribution may be well approximated by the normal distribution having the same mean and variance, i.e. N(500, 250). Of course, Bi(1000, 0.5) is discrete, whereas N(500, 250) is continuous. However, when talking about $P(X < 520|H)$ the continuous approximation will still give a sensible probability. The only concession we need make to the discreteness of Bi(1000, 0.5) is to rewrite the required probability as $P(X \le 519.5|H)$ since a continuous approximation is generally best when taken at values between jumps in the discrete distribution function to which it is approximating.

{...........

Examples

In the above example, the correct probability is found to be

$$P(X < 520 | H) = 0.891276 . \qquad (7.68)$$

This computation involved summing 520 individual probabilities, each very small, and would be impossible in practice without a computer. In contrast, the normal approximation is

$$P(X < 520 | H) \approx P(N(500, 250) < 519.5)$$

$$= P(N(0, 1) \leq (519.5 - 500) / 250^{\frac{1}{2}})$$

$$= \Phi(1.2333) .$$

We can see immediately from Table 7.1 that this probability is between $\Phi(1) = 0.8413$ and $\Phi(1.5) = 0.9332$, and this may be good enough for many purposes. From more extensive tables we find $\Phi(1.233) = 0.8913$, which is very close to (7.68) and required only trivial calculations.

As a second example, consider $P(55 \leq Po(60) \leq 65)$, in which m is not particularly large. The correct answer, 0.52235, required quite heavy calculation. The normal approximation

$$P(54.5 \leq N(60, 60) \leq 65.5) = \Phi(0.710) - \Phi(-0.710) = 0.5222$$

(using a book of tables) is very accurate again.

...........}

Other relationships between distributions may be found which are exact, rather than depend on taking limits. Exercise 7(c)6 is an example.

Several important theorems concern two random variables, using Bayes' theorem and extending the argument.

The normal-normal theorem

Let $X | H \sim N(m, v)$ and $Y | (X=x) \wedge H \sim N(x, w)$. Then

$$Y | H \sim N(m, v+w) , \qquad (7.69)$$

$$X | (Y=y) \wedge H$$

$$\sim N((v+w)^{-1}(vy + wm), (v+w)^{-1}vw) . \qquad (7.70)$$

{...........}

Proof

$$d_{X,Y|H}(x,y) = \left[(2\pi v)^{-\frac{1}{2}} \exp\{-(x-m)^2/(2v)\} \right] \times$$
$$\times \left[(2\pi w)^{-\frac{1}{2}} \exp\{-(y-x)^2/(2w)\} \right]$$
$$= (2\pi)^{-1}(vw)^{-\frac{1}{2}} \exp(-r/2), \tag{7.71}$$

where

$$r = v^{-1}(x-m)^2 + w^{-1}(y-x)^2$$
$$= (v^{-1}+w^{-1})x^2 - 2(v^{-1}m+w^{-1}y)x + (v^{-1}m^2+w^{-1}y^2)$$
$$= (v^{-1}+w^{-1})\{x - (v^{-1}+w^{-1})^{-1}(v^{-1}m+w^{-1}y)\}^2 + q,$$
$$q = v^{-1}m^2 + w^{-1}y^2 - (v^{-1}+w^{-1})^{-1}(v^{-1}m+w^{-1}y)^2$$
$$= (v+w)^{-1}(y-m)^2.$$

Therefore, integrating (7.71) with respect to x (which corresponds to extending the argument as in equation (7.65)), we find

$$d_{Y|H}(y)$$
$$= (2\pi)^{-1}(vw)^{-\frac{1}{2}} \exp\{-(v+w)^{-1}(y-m)^2/2\} \{2\pi(v^{-1}+w^{-1})\}^{\frac{1}{2}}$$
$$= \{2\pi(v+w)\}^{-\frac{1}{2}} \exp\{-(v+w)^{-1}(y-m)^2/2\} \tag{7.72}$$

which proves (7.69). Now applying Bayes' theorem as in equation (7.46) entails dividing (7.71) by (7.72). The result (7.70) now follows easily.
...........}

An example of this theorem has already been given in Section 7.7, the heat transfer example. To obtain the results of that section, let $X=C$, $Y=M$, $m=9$, $v=1$, $x=0.25$ and $y=8.4$. The scope of application of this theorem is very wide because the heat transfer example is typical of a great many practical problems. We are often interested in quantities which we can only measure experimentally, and where there is significant measurement error. Denoting X as the quantity of interest and Y as the experimental measurement, the belief that $Y|(X=x)\wedge H \sim N(x,w)$ is reasonable. It says in particular that Y is an unbiased measurement, neither tending to be too low or too high, because $E(Y|(X=x)\wedge H)=x$. However, the variance w represents experimental error. Prior knowledge of the quantity of interest is expressed in $X|H \sim N(m,v)$. After the experiment, your estimate of X becomes the posterior mean

$$E(X|(Y=y)\wedge H) = (v+w)^{-1}(vy+wm),$$

which is a combination of the experimental value y and the prior mean m. In Chapter 9 we shall extend this approach to the case of several measurements.

The beta-binomial theorem

Let $X|H \sim Be(p,q)$ and $Y|(X=x) \wedge H \sim Bi(n,x)$. Then

$$P(Y=y|H) = \binom{n}{y} B(p+y, q+n-y)/B(p,q),\qquad (7.73)$$

$$X|(Y=y) \wedge H \sim Be(p+y, q+n-y).\qquad (7.74)$$

{...........

Proof

Extending the argument using (7.62), we find the joint semi-density function from (7.59):

$$d_{X,(Y=y)|H}(x)$$

$$= \left[\{B(p,q)\}^{-1} x^{p-1}(1-x)^{q-1} \right] \left[\binom{n}{y} x^y (1-x)^{n-y} \right]$$

$$= \{B(p,q)\}^{-1} \binom{n}{y} x^{p+y-1}(1-x)^{q+n-y-1},$$

for $0 \le x \le 1$. Therefore

$$P(Y=y|H) = \{B(p,q)\}^{-1} \binom{n}{y} \int_0^1 x^{p+y-1}(1-x)^{q+n-y-1} \, dx$$

which gives (7.73) immediately. Applying Bayes' theorem in the form (7.65) gives (7.74).

...........}

The family of distributions (7.73) are standard distributions that we have not mentioned previously. Because of this theorem, they are known as beta-binomial distributions. The simplest way to find their mean and variance is by the expectation of an expectation theorem – (6.30) and (6.31). For instance, $E(Y|X \wedge H) = n X$, $\therefore E(Y|H) = nE(X|H) = np/(p+q)$. Applications of the beta-binomial theorem are also important, and are discussed fully in Chapter 8.

The Poisson-gamma theorem

Let $X|H \sim Ga(a,b)$ and $Y|(X=x) \wedge H \sim Po(x)$. Then

$$P(Y=y|H) = \left(\frac{\Gamma(b+y)}{\Gamma(b)y!} \right) \left(\frac{a}{a+1} \right)^b \left(\frac{1}{a+1} \right)^y,\qquad (7.75)$$

$$X|(Y=y) \wedge H \sim Ga(a+1, b+y).\qquad (7.76)$$

The proof of this theorem appears as Exercise 7(f)3 below. Notice that if b is an integer, then (7.75) is the $NB(b, a/(a+1))$ distribution. We can therefore think of it as extending the negative-binomial family so that its first parameter can be any positive number, not necessarily an integer. The mean and variance remain as given in Section 5.8.

{............

Example

Research into industrial accidents shows that some people consistently have more accidents than others. They are said to be more accident prone. If a person's accident rate is known then it is reasonable to suppose that his accidents occur randomly in time with that rate. Therefore, if R is the accident rate, in accidents per year, for a new employee at a factory, and if A is the number of accidents this worker will actually have in his first year, then

$$A \mid (R=r) \wedge H \sim \text{Po}(r)$$

is an objective judgement. Using his experience (contained in H), the employer measures his distribution for $R \mid H$ as follows. He judges that the average number of accidents per worker per year is 1.5, and he assigns

$$E_1(R \mid H) = 1.5 .$$

He believes that the majority of workers will have accident rates of between 0.5 and 3 per year. Regarding this range as comprising plus and minus one standard deviation he has

$$var_1(R \mid H) = (2.5/2)^2 = 1.5625 .$$

His distribution is unimodal. He adopts a gamma distribution and chooses its parameters to satisfy

$$b/a = 1.5 , \qquad\qquad b/a^2 = 1.5625$$

$$\therefore a = 1.5/1.5625 \approx 1 \qquad \therefore b \approx 1.5 .$$

His distribution is therefore

$$R \mid H \quad_1\!\sim \text{Ga}(1, 1.5) .$$

Applying the Poisson-gamma theorem,

$$R \mid (A=a) \wedge H \quad_1\!\sim \text{Ga}(2, 1.5+a) .$$

He actually observes that this worker has no accidents in his first year. Therefore his distribution for R is now measured as

$$R \mid (A=0) \wedge H \quad_1\!\sim \text{Ga}(2, 1.5) .$$

In particular, his expectation of this worker's accident rate is $1.5/2 = 0.75$. His variance is $1.5/4 = 0.375$, giving a standard deviation of 0.61. He therefore has a probability of about 0.65 that this worker's accident rate is between $0.75-0.61=0.14$ and $0.75+0.61=1.36$ accidents per year. He clearly has quite a high probability that his accident rate is lower than the average rate of 1.5.

............}

Exercises 7(f)

1. Let X be the number of successes in 625 binary trials with probability 0.1 of success in each trial, given information H. Calculate an approximate value for $P(X>77|H)$.

2. Let T be the freezing point (in °C) of a liquid proposed for use in polar regions. Its suppliers claim that $T=-38$, but You doubt this and measure $T|H \sim N(-36,16)$, so that T is probably between -39 and -23. A freezing point as high as -25°C would be unsatisfactory for Your purposes, so You perform Your own measurement of T. You judge that the inaccuracy of Your measurement X is represented by a standard deviation of 3. Given the actual measurement $X=-35$, what should You now believe about T? In particular, approximately how large is $P(T>-25|H \wedge (X=-35))$?

3. Prove the Poisson-gamma theorem.

8
Frequencies

Particularly in the social sciences, the most common form of statistical data is a frequency, i.e. the number of propositions of a certain kind which are observed to be true. Examples are the number of houses in a street which are inhabited by six or more people, or the number of people in a sample who say they will vote for a particular political party. In our study of trials and deals we considered two special structures for frequencies. This chapter introduces a much more general structure, called exchangeability. Exchangeable sequences of propositions are defined in Section 8.1. Sections 8.2 and 8.3 present a measurement device, by extending the argument, for probabilities in exchangeable sequences. Section 8.2 deals with finite sequences, whereas the more important case of infinite exchangeable sequences is dealt with in Section 8.3.

With frequency data, the most important questions concern the unobserved propositions in the sequence. For instance, given the frequencies of voting intentions in a sample, what are the likely voting intentions of voters who were not in the sample, and hence what result might be expected if an election were held? In Section 8.4, posterior probabilities for some or all of the unobserved propositions are derived, and particularly simple formulae arise in Section 8.5 when the prior distribution for the proportion of true propositions is a member of the beta family. Section 8.6 considers the effect of observing a large number of propositions. For instance, if 31% of a large sample of voters say they will vote Liberal, then it is natural to suppose that about 31% of all voters would vote Liberal. The frequency notion of probability is explored, and the difficulty of basing a theory of probability entirely on a frequency definition is explained. These ideas are continued in Section 8.7, which considers what you should do if your measured probabilities conflict with observed frequencies.

8.1 Exchangeable propositions

An agricultural botanist is breeding a strain of onions which is resistant to the onion fly. She plants 50 seeds of her latest strain. Let $F_i \equiv$ 'onion plant will be attacked by the onion fly', $i = 1, 2, \ldots, 50$. She is interested in expressing probabilities for the F_is, and in particular the distribution of the random variable $X \equiv$ the number of onion plants attacked. Now her probabilities for the F_is will have an important property called *exchangeability*. Since she has no knowledge concerning individual seeds, which would cause her to give different probabilities to different F_is, she will give them all the same probability,

$$P(F_i|H) = P(F_1|H), \quad i = 2, 3, \ldots, 50. \tag{8.1}$$

Furthermore,

$$P(F_i \wedge F_j | H) = P(F_1 \wedge F_2 | H),\qquad\qquad(8.2)$$

for all $i \neq j$. Indeed, her probability that any K specified plants will be attacked is the same as for any other K specified plants, and in particular equals $P(\wedge_{i=1}^{k} F_i | H)$.

Exchangeability is a very common property. We frequently wish to consider a set of similar propositions such that we have no information relating to specific propositions.

{............

Examples

It would be reasonable to consider each of the following as defining a set of exchangeable propositions.

$H_i \equiv$ 'heads on the i-th toss of a coin'.

$D_i \equiv$ 'i-th child in a school class will die before age 40'.

$F_i \equiv$ 'i-th car passing a road junction will be a Ford'.

$B_i \equiv$ 'i-th patient with bronchitis will develop pneumonia'.

$S_i \equiv$ 'i-th sample of an alloy will break under a certain stress'.

The criterion for exchangeability is that your information must be the same about all propositions or groups of propositions. You might not regard the D_is as exchangeable if you know the medical history of any of the children or their parents. The first test of exchangeability, (8.1), would fail in this case. The F_is could perhaps fail the second test, (8.2), if you thought that Fords might tend to travel together.

............}

Exchangeable propositions

The propositions $E_1, E_2, \ldots, E_n$ are said to be exchangeable given H if for all $m = 1, 2, \ldots, n$ and for every set of m different integers $j_i, j_2, \ldots, j_m$,

$$P(\wedge_{i=1}^{m} E_{j_i} | H) = P(\wedge_{i=1}^{m} E_i | H).\qquad\qquad(8.3)$$

Two special cases of exchangeable propositions have been discussed in Chapter 4. Trials are exchangeable because, if the E_is are trials given H with $P(E_i | H) = p$ then both sides of (8.3) equal p^m. Propositions which are exchangeable and independent are trials. The propositions in a deal are also exchangeable, as the exchangeability of deals theorem in Section 4.9 shows. However, neither of these cases would be appropriate for the botanist. If the F_is were trials then

$$P(F_2|F_1 \wedge H) = P(F_2|H),$$
(8.4)

and if they were a deal then

$$P(F_2|F_1 \wedge H) < P(F_2|H).$$
(8.5)

Learning that one plant has been attacked by onion fly would change her probability that another would be attacked, because it would change her belief about how resistant the strain was. Therefore (8.4) is inappropriate. Also. each plant found to be attacked would diminish her confidence in the resistant property, and therefore increase her probability for another being attacked. Instead of (8.5), her probabilities would satisfy

$$P(F_2|F_1 \wedge H) > P(F_2|H).$$
(8.6)

Dependence of the form (8.6) is most commonly found with exchangeable propositions.

The judgement of exchangeability is generally objective, but the assignment of particular probabilities is still typically subjective. Almost all people would agree for instance with (8.1) in the onion example, but different botanists might give different values to $P(F_1|H)$. Because exchangeability equates so many probabilities, only n probabilities need to be measured in order to imply probabilities for all combinations of a set of n exchangeable propositions. Define

$$q_m = P(\wedge_{i=1}^{m} E_i|H), \qquad m=1,2,\ldots,n,$$
(8.7)

then any probability concerning the E_is can be expressed in terms of the q_ms. For instance

$$P(E_2 \wedge (\neg E_6)|H) = P(E_2|H) - P(E_2 \wedge E_6|H) = q_1 - q_2,$$

or

$$P(E_1 \wedge (\neg E_7) \wedge E_8 \wedge (\neg E_{10})|H)$$
$$= P(E_1 \wedge (\neg E_7) \wedge E_8|H) - P(E_1 \wedge (\neg E_7) \wedge E_8 \wedge E_{10}|H)$$
$$= P(E_1 \wedge E_8|H) - P(E_1 \wedge E_7 \wedge E_8|H) - P(E_1 \wedge E_8 \wedge E_{10}|H)$$
$$+ P(E_1 \wedge E_7 \wedge E_8 \wedge E_{10}|H)$$
$$= q_2 - 2q_3 + q_4.$$

One natural consequence is that the symmetry of the propositions extends beyond (8.3). The probability given H of any specified E_is being true and any other specified E_is being false depends only on the numbers of true and false E_is specified, and not on which ones are specified.

Exchangeability and symmetry

If $E_1, E_2, \ldots, E_n$ are exchangeable given H, then the probability of any proposition concerning the E_is remains the same if the E_is are shuffled into a different order. In particular, if F asserts that r specified E_is are true and another s specified E_is are false then

$$P(F|H) = P((\wedge_{i=1}^{r} E_i) \wedge (\wedge_{i=r+1}^{r+s} \neg E_i)|H). \tag{8.8}$$

The proof of (8.8) is implicit in the proof of the following result.

Characterization by q_ms

Let $E_1, E_2, \ldots, E_n$ be exchangeable given H and let $q_1, q_2, \ldots, q_n$ be defined by (8.7). Let X be the number of propositions in any m specified E_is which are true. Then the distribution of X given H is

$$P(X=x|H) = \binom{m}{x} \sum_{j=0}^{m-x} \binom{m-x}{j} (-1)^j q_{x+j}. \tag{8.9}$$

{...........

Proof
Let F be any proposition asserting that, of the m specified E_is, a certain specified x are true and the remainder false. Then (8.9) will follow immediately if we can demonstrate that

$$P(F|H) = \sum_{j=0}^{m-x} \binom{m-x}{j} (-1)^j q_{x+j}, \tag{8.10}$$

since there are $\binom{m}{x}$ such propositions F implied by '$X=x$'.

We prove (8.10) by induction. First note that it is true for any m when $x=m$, giving the correct probability q_m. Suppose that (8.10) is true for $m-x < y$ and we will prove it for $m-x = y$. Let E_k be one of the $m-x = y$ propositions asserted by F to be false. Then

$$F = G \wedge (\neg E_k),$$

where G is a proposition asserting that a specified x propositions are true and another $m-x-1 = y-1$ are false. We have

$$G = F \vee (G \wedge E_k).$$

Since $G \wedge E_k$ asserts that a specified $x+1$ propositions are true and another specified $m-x-1 = y-1$ are false, we can apply (8.10) (which is assumed true for $m-x = y-1$) to both F and $G \wedge E_k$, to give

$$P(F|H) = P(G|H) - P(G \wedge E_k|H)$$

$$= \sum_{j=0}^{m-x-1} \binom{m-x-1}{j}(-1)^j q_{x+j}$$

$$- \sum_{j=0}^{m-x-1} \binom{m-x-1}{j}(-1)^j q_{x+j+1}$$

$$= \sum_{j=0}^{m-x} a_j q_{x+j} \, ,$$

where

$$a_j = \binom{m-x-1}{j}(-1)^j - \binom{m-x-1}{j-1}(-1)^{j-1}$$

$$= (-1)^j \frac{(m-x-1)!}{j!(m-x-j)!} \{(m-x-j)+j\}$$

$$= (-1)^j \binom{m-x}{j} .$$

............}

Unfortunately, elaboration such as (8.9), using the q_ms, is not good elaboration. Although all probabilities of interest may be elaborated in terms of them, the q_ms are not easy to measure accurately. It is obvious that $q_m \le q_{m-1}$, but how much smaller should it be? The difficulties of measurement are compounded by the fact that there are a whole series of inequalities which the q_ms must satisfy. The simplest of these are

$$q_2 \le q_1 \le (q_2+1)/2 . \tag{8.11}$$

The first part of (8.11), that $q_2 \le q_1$, is obvious and implied by the conjunction inequality. To prove the second part, we begin with the disjunction theorem.

$$P(E_1 \vee E_2|H) = P(E_1|H) + P(E_2|H) - P(E_1 \wedge E_2|H)$$

$$= 2q_1 - q_2$$

since the E_is are exchangeable. The fact that $P(E_1 \vee E_2|H) \le 1$ proves the result.

In Section 8.2 we offer a superior elaboration.

8.2 The finite characterization

Suppose You know that precisely r of the E_is are true, but not which ones are true. Let this information be H_r. All selections of r from the n E_is are equally likely to comprise the r true propositions (because they are exchangeable). Then the E_is are a complete deal of n from $(r, n-r)$, and this structure provides all Your probabilities. In particular,

$$q_m = P(\wedge_{i=1}^m E_i|H_r) = \frac{r(r-1)\cdots(r-m+1)}{n(n-1)\cdots(n-m+1)} \, , \tag{8.12}$$

and if X is the number of true propositions in any specified m E_is,

$$X|H \sim \text{Hy}(n,r,m).$$ (8.13)

More realistically, You will not know how many of the n E_is are true. Let the random variable R be the number of true propositions in the set. Then the information H_r represents Your current information H (in which R is unknown) plus the knowledge that $R=r$, i.e.

$$H_r = H \wedge (R=r).$$

Therefore we can elaborate by extending the argument to include R. For any proposition F,

$$P(F|H) = \sum_{r=0}^{n} P(F|H \wedge (R=r)) P(R=r|H).$$ (8.14)

The judgement that the E_is form a complete deal of n from $(r, n-r)$ given $H \wedge (R=r)$ is objective, relying only on the judgement of exchangeability. Therefore the components $P(F|H \wedge (R=r))$ in (8.14) will be objective. The remaining components, $P(R=r|H)$, form Your subjective distribution for R given H. The following theorem gives versions of (8.14) for the most useful forms of F, using (8.12) and (8.13).

The finite characterization theorem
Let $E_1, E_2, \ldots, E_n$ be exchangeable given H. Let R be the number of E_is which are true. Then, with q_m defined by (8.7),

$$q_m = \sum_{r=0}^{n} \frac{r(r-1) \cdots (r-m+1)}{n(n-1) \cdots (n-m+1)} P(R=r|H).$$ (8.15)

If X is the number of true propositions in a specified m E_is, then

$$P(X=x|H) = \sum_{r=0}^{n} \frac{\binom{r}{x} \binom{n-r}{m-x}}{\binom{n}{m}} P(R=r|H).$$ (8.16)

These elaborations, like those of Section 8.1 using the q_ms, require essentially n probabilities to be measured subjectively. Strictly, there are $n+1$ probabilities $P(R=r|H)$ for $r=0, 1, \ldots, n$, but they form a distribution and so must sum to one. Therefore, as soon as n of them have been measured the last will follow automatically. In practice it is much easier to measure Your distribution for R accurately than to measure the q_ms accurately.

{...........

Example

The analysis of the job applications example in Section 5.5 uses this characterization. There were $n = 6$ propositions, $E_i =$ 'the i-th application is from a woman'. In Section 5.5 we denoted the total number of women applicants by Z, rather than R. A measured distribution was provided for Z, from which other probabilities were derived, using (8.16).

...........}

The distribution of R given H may be measured by the techniques of Chapter 5, except that none of the standard distributions given in Section 5.8 would be very suitable for this purpose. For, suppose that Your measured distribution for R given H is $\text{Bi}(n,p)$. It is easy to demonstrate that then the E_is are trials given H, which we have already remarked is not usually appropriate. If $R|H \sim \text{Hy}(N,R,n)$ then it is equally easy to show that $X|H \sim \text{Hy}(N,R,m)$, which is even less suitable in practice. Negative binomial and Poisson distributions are strictly only for unbounded random variables. It is therefore fortunate that in practice we rarely need to use the finite characterization theorem. To understand why this is the case we return to the example of the onions.

To use the finite characterization theorem the botanist must give her distribution for R, the total number of plants which will be attacked. The possible values of R are $0, 1, \ldots, 50$ because she has planted 50 seeds. She could have planted 100 or 1000 seeds, and in each case a different distribution would be required. The propositions $F_1, F_2, \ldots, F_{50}$ are really part of a much larger exchangeable set. She will only observe the first 50 but the others are of interest. The success of the new onion strain will depend on what proportion of all future seeds grow into plants susceptible to attack from the onion fly. It is therefore important that she see $F_1, F_2, \ldots, F_{50}$ as embedded in a larger, effectively infinite, exchangeable set. We shall see in Sections 8.3 to 8.5 that it is actually easier to work with an infinite exchangeable sequence than a finite one.

It is nearly always realistic and relevant to consider an infinite exchangeable sequence. On the rare occasions when an exchangeable set is necessarily finite, n is usually small enough for $P(R = r|H)$ to be measured directly.

Exercises 8(a)

1. In Exercise 4(c)3 it would perhaps be more reasonable to suppose that the propositions R_j are exchangeable rather than independent. In this case, how would you elaborate your probability that six or more rats will show positive responses?

2. Would you regard the propositions in Exercise 4(a)3 as exchangeable?

3. An auditor examines the expenses invoices of a company. Letting $V_i \equiv$ 'i-th invoice is validated', he regards the V_is as exchangeable given his information H. He measures $q_1 = 0.95$, $q_2 = 0.92$ and $q_3 = 0.90$. Measure his probabilities $P(V_5 \wedge V_6 \wedge ((\neg V_7) \vee (\neg V_8)) | H)$ and $P(V_3 | V_1 \wedge (\neg V_2) \wedge H)$.

4. Consider the sequence $R_1, R_2, \ldots, R_n$, where R_i is the proposition of a red ball on the i-th draw from Polya's urn (Exercise 1(c)2). H denotes your information before any draws are made. Let E be a proposition asserting that a specified subset of the n R_is are true and the remainder false. Show that $P(E|H)$ depends on the number, m, of propositions in the subset asserted to be true, but not on which ones are specified.

5. Exercise 4 proves that the R_is in Polya's urn are exchangeable given H. Find the probabilities $q_1, q_2, \ldots, q_n$, and the distribution given H of $M \equiv$ the number of true propositions in the whole sequence of n.

6. A school dentist examines the teeth of a class of 32 children. If $T \equiv$ the number of children whose teeth will be found to need urgent attention, her distribution for T is as follows.

t	0	1	2	3	4
$P_d(T=t|H)$	0.25	0.40	0.20	0.10	0.05

and $P_d(T > 4 | H) = 0$. She will examine 12 children in one session. Let X be the number of children in this session whose teeth need urgent attention, and compute her distribution for X given H.

8.3 De Finetti's theorem

An infinite exchangeable sequence is defined formally as follows.

Infinite exchangeable sequence
Let $E_1, E_2, E_3, \ldots$ be a sequence of propositions. They are said to be exchangeable given H if for every $n = 2, 3, \ldots$ the propositions $E_1, E_2, \ldots, E_n$ are exchangeable given H.

This definition ensures that equation (8.3) holds for arbitrarily large m.

Now consider what happens to the finite characterization theorem as $n \to \infty$. The first point to note is that we can no longer extend the argument to R because R will also tend to infinity. Instead of working in terms of the number of true propositions we use the proportion. So if R_n is the number of true propositions in the first n, define

$$T \equiv \lim_{n \to \infty} \frac{R_n}{n}.$$

T is the proportion of true propositions in the infinite sequence. Often, T is the random variable of primary interest. In the onion example, for instance, T represents the proportion of onions that will be attacked, out of all onions ever grown from this strain. It is an overall measure of susceptibility for the strain.

Now consider the E_is conditional on $T=t$. The distribution of X given $H \wedge (R_n=r_n)$ is $\text{Hy}(n, r_n, m)$ and letting both n and r_n tend to infinity with the proportion $r_n/n=t$ fixed we have

$$X|(T=t) \wedge H \sim \text{Bi}(m, t), \tag{8.17}$$

(see Section 4.11). Therefore, conditional on $(T=t) \wedge H$, the E_is become trials with $P(E_i|(T=t) \wedge H)=t$.

It remains to consider what happens to the distribution of R_n/n given H as $n \to \infty$. The limiting distribution function could take any form, but as usual the important cases are when it is a step function or a differentiable function. Your information about T will generally be such that T is continuous with all numbers in the range $[0, 1]$ as possible values. Occasionally, however, T may be discrete, with only specific possible values. To deal with these two cases we require two different formulae for extending the argument. For discrete T,

$$P(X=x|H) = \sum_t P(X=x|(T=t) \wedge H) P(T=t|H)$$

$$= \binom{m}{x} \sum_t t^x (1-t)^{m-x} P(T=t|H), \tag{8.18}$$

the sum being over all possible values t of T. For continuous T,

$$P(X=x|H) = \int_0^1 P(X=x|(T=t) \wedge H) \, d_{T|H}(t) \, dt$$

$$= \binom{m}{x} \int_0^1 t^x (1-t)^{m-x} \, d_{T|H}(t) \, dt. \tag{8.19}$$

Using the corresponding definitions of expectation, both forms can be covered by a single general statement.

De Finetti's theorem

Let $E_1, E_2, E_3, \ldots$ be an infinite exchangeable sequence given H, and let T be the proportion of true propositions in the sequence. Then, with q_m defined by (8.7),

$$q_m = E(T^m|H), \tag{8.20}$$

and if X is the number of true propositions in a specified m E_is

$$P(X=x|H) = \binom{m}{x} E(T^x (1-T)^{m-x}|H). \tag{8.21}$$

The distribution of T given H can be measured in the usual way. If there are only a few possible values for T then the discrete probabilities $P(T=t|H)$ should be measured directly. We shall see an example of this in Section 8.4. By far the most common case is when T is continuous on the range $[0, 1]$, and in this case it is appropriate to choose a distribution from the beta family. This approach is detailed in Section 8.5.

8.4 Updating

Consider an infinite sequence $E_1, E_2, E_3, \ldots$, exchangeable given H, and suppose that You observe that eight of the first 50 propositions are true. Your information is now $(X=8) \wedge H$, defining X to be the number of true propositions in the first 50. What are Your beliefs about the remainder of the sequence, $E_{51}, E_{52}, \ldots$? For example, the botanist might observe that eight out of the 50 onion plants are damaged by the onion fly: what does this tell her about this onion strain? More generally, let X be the number of true propositions in the first (or any other) m E_is: what probabilities should $E_{m+1}, E_{m+2}, \ldots$ have given information $(X=x) \wedge H$?

Two facts are simple but important.

The conditional exchangeability theorem

(1) Let $E_1, E_2, E_3, \ldots$ be exchangeable given H, and let F be any proposition asserting that some specified E_is are true and other specified E_is are false. Then those E_is which are not specified by F are exchangeable given $F \wedge H$.

(2) Let $F_1, F_2, \ldots, F_b$ be a partition given H, and let $E_1, E_2, E_3, \ldots$ be exchangeable given $F_c \wedge H$ for each $c = 1, 2, \ldots, b$. Then $E_1, E_2, E_3, \ldots$ are exchangeable given H.

{............

Proof

Part (1) is proved by Bayes' theorem. Suppose that F asserts that r_1 specified propositions are true and s are false. Let G be any proposition asserting that r_2 specified E_is are true, none of them included in those specified in F. Then

$$P(G|F \wedge H) = P(F \wedge G|H)/P(F|H).$$

Now $F \wedge G$ asserts that a specified r_1+r_2 E_is are true and another s are false. Its probability is given by (8.10) with $x = r_1+r_2$ and $m = r_1+r_2+s$, and does not depend on which r_2 propositions are specified by G. Therefore, $P(G|F \wedge H)$ depends only on the number, r_2, of propositions specified by G, and not on which ones are specified.

Part (2) is proved by extending the argument. Let

$$q_{m,c} = P(\wedge_{i=1}^m E_i | F_c \wedge H) = P(\wedge_{i=1}^m E_{j_i} | F_c \wedge H)$$

for all selections of m different numbers $j_1, j_2, \ldots, j_m$. Then

$$P(\wedge_{i=1}^m E_{j_i} | H) = \sum_{i=1}^b q_{m,c} P(F_c | H) = P(\wedge_{i=1}^m E_i | H).$$

..........}

The above formal proof looks complicated but the theorem is obvious when we think of exchangeability as being symmetry of belief. In part (1), H contains no information which relates to any specific group of E_is. Clearly, learning F gives no information relating to any specific E_is except those referred to in F itself. The others must remain exchangeable. In part (2), even if You knew which F_c in the partition was true You would have no information relating to specific E_is, so it is clear that You have no such information to begin with.

We can now assert that $E_{m+1}, E_{m+2}, \ldots$ will be exchangeable given $(X=x) \wedge H$. The result may be proved formally using the two parts of the conditional exchangeability theorem, but it is also obvious from symmetry considerations.

Just as Your exchangeable probabilities for the E_is given H are characterized by Your distribution for T given H, through de Finetti's theorem, Your exchangeable distribution after observing $X=x$ is characterized by Your distribution for T given $(X=x) \wedge H$. This is easily obtained by Bayes' theorem. Then de Finetti's theorem may be used to give the required probabilities. These operations are all straightforward, and lead directly to the following theorem.

Exchangeable updating

Let $E_1, E_2, E_3, \ldots$ be exchangeable given H. Let X be the number of true propositions in the first m E_is. Let T be the proportion of true propositions in the sequence. Then

(1) if T is discrete,

$$P(T=t | (X=x) \wedge H)$$

$$= \frac{P(T=t|H) t^x (1-t)^{m-x}}{\sum_u P(T=u|H) u^x (1-u)^{m-x}}, \qquad (8.22)$$

where the sum is over all possible values u of T;

(2) if T is continuous,

$$d_{T|(X=x) \wedge H}(t) = \frac{t^x (1-t)^{m-x} d_{T|H}(t)}{\int_0^1 u^x (1-u)^{m-x} d_{T|H}(u) \, du}; \qquad (8.23)$$

(3) the equivalent of q_m after information $X=x$ is

$$q_k^* \equiv P(\wedge_{i=m+1}^{m+k} E_i | (X=x) \wedge H)$$

$$= E(T^k | (X=x) \wedge H), \tag{8.24}$$

and for $i = m+1, m+2, \ldots$

$$P(E_i | (X=x) \wedge H) = q_1^* = E(T | (X=x) \wedge H); \tag{8.25}$$

(4) if Y is the number of true propositions in $E_{m+1}, \ldots, E_{m+r}$ then

$$P(Y=y | (X=X) \wedge H)$$

$$= \binom{r}{y} E(T^y (1-T)^{r-y} | (X=x) \wedge H). \tag{8.26}$$

To illustrate the computations involved in these formulae, we consider a practical application in which T is discrete. A certain flower has been bred in two varieties. One variety has pink flowers and the other has white flowers. If seed is produced by cross-fertilizing one flower of each variety, then some offspring will have pink flowers and others white. From genetic theory You believe that either three-quarters will be pink or three-quarters will be white (because one colour is 'dominant'). This does not mean that if four seeds were germinated then You will either have three white and one pink or three pink and one white. You could also have two of each, or four the same colour. The statement refers to the proportions in an infinite sequence of seeds. So let $W_i \equiv$ 'i-th plant is white'. You judge the W_is to be exchangeable given Your information H. Letting T be the proportion of white flowers in the sequence, You know that $T=1/4$ or $T=3/4$. You have no prior information to suggest that one alternative is more probable than the other, so

$$P(T=1/4 | H) = 1/2, \qquad P(T=3/4 | H) = 1/2.$$

The symmetry of Your information is such that all these measurements are essentially true values.

You now obtain six plants from cross-fertilized seed and observe that five have white flowers. Your information is now $(X=5) \wedge H$ with $m=6$. It is now obviously far more likely that $T=3/4$ than $T=1/4$. The updating theorem allows us to measure accurately just how much more likely it is. We apply equation (8.22) to obtain Your posterior beliefs about T.

$$P(T=1/4 | (X=5) \wedge H) = \frac{0.5 \times 0.25^5 \times 0.75}{(0.5 \times 0.25^5 \times 0.75) + (0.5 \times 0.75^5 \times 0.25)}$$

$$= 0.012.$$

Similarly, $P(T=3/4 | (X=5) \wedge H) = 0.988$. The proposition that $T=1/4$ (pink flowers are 'dominant') is now very improbable.

If You were now to plant a further six cross-bred seeds, Your probability for any one of them being white is, from (8.25),

$$P(W_i|(X{=}5){\wedge}H) = E(T|(X{=}5){\wedge}H)$$

$$= (0.25 \times 0.012) + (0.75 \times 0.988) = 0.744 ,$$

for $i{=}7,8,\dots,12$. Your distribution for $Y \equiv$ the number of white flowers in the second group of six may be obtained from (8.26). For instance, Your probability of repeating the first experiment exactly is

$$P(Y{=}5|(X{=}5){\wedge}H) = \binom{6}{5} E(T^5(1{-}T)^1|(X{=}5){\wedge}H)$$

$$= 6 (0.25^5 \times 0.75 \times 0.012 + 0.75^5 \times 0.25 \times 0.988) = 0.352 .$$

Exercises 8(b)

1. Consider Polya's urn, as in Exercise 1(c)2 or 8(a)4-5, but the urn initially contains a red balls and b white balls instead of one of each. With R_i denoting a red ball on the i-th draw as usual, show that the R_is are exchangeable. [Hint: suppose that this is the original urn after $a{-}1$ red balls and $b{-}1$ white balls have been drawn.]

2. Let $E_1,E_2,E_3,\dots$ be an infinite sequence of propositions, exchangeable given H. Using de Finetti's theorem, prove that $q_2 \geq q_1^2$, and therefore that

$$P(E_i|H{\wedge}E_j) \geq P(E_i|H) .$$

3. If an atomic particle of type A is struck by a particle of type B then a particle of type P may be produced. It has not been possible to perform this experiment yet, but on theoretical grounds a physicist believes that either (a) a P-particle will always be produced, or (b) given a long sequence of experiments P-particles will be produced in just half of the experiments. Let E be the proposition that P-particles are always produced. If the physicist were given the equipment to perform the experiment, describe how his probability for E should change after the first three experiments.

4. You are a keen observer of birds. One day you see a bird having a small white bar under the wing, a mark which you had not noticed before in birds of this species. You entertain three theories about the white bar – $T_1 \equiv$'all birds of this species have it', $T_2 \equiv$'it is borne by either males or females but not both', $T_3 \equiv$'it is a result of a mutation or accident to this bird, and no others will have it'. You regard these as an equi-probable partition. You now closely observe ten more birds of this species and they all have the white bar. Measure your probability for T_1.

8.5 Beta prior distributions

When T is continuous the natural choice of family to measure its distribution is the beta family, since these are all confined to possible values in the range $[0,1]$ and offer a wide variety of shapes. Furthermore, the calculations required by equations (8.23) to (8.26) are extremely simple if $T|H$ has a beta distribution.

Updating with beta prior distribution

Let $E_1, E_2, E_3, \ldots$ be exchangeable given H, and let T be the proportion of true propositions in the sequence. Let X be the number of true propositions in $E_1, E_2, \ldots, E_m$. If

$$T|H \sim Be(a,b)$$

then

(1) $T|(X=x) \wedge H \sim Be(a+x, b+m-x)$, \hfill (8.27)

(2) $q_k^* = \dfrac{a+x}{a+b+m} \dfrac{a+x+1}{a+b+m+1} \times \cdots$

$$\times \frac{a+x+k-1}{a+b+m+k-1} ,$$ \hfill (8.28)

where q_k^* is defined by (8.24),

(3) $P(E_i|(X=x) \wedge H) = \dfrac{a+x}{a+b+m}$, \hfill (8.29)

for $i = m+1, m+2, \ldots$, and

(4) $P(Y=y|(X=x) \wedge H)$

$$= \binom{r}{y} \frac{B(a+x+y, b+m+r-x-y)}{B(a+x, b+m-x)} ,$$ \hfill (8.30)

where Y is the number of true propositions in the sequence $E_{m+1}, E_{m+2}, \ldots, E_{m+r}$.

{...........}

Proof

(1) Equation (8.27) follows directly from the beta-binomial theorem, equation (7.74).

(2) By (8.24) and (8.27)

$$q_k^* = \int_0^1 t^k \, t^{a+x-1} (1-t)^{b+m-x-1} \, dt \, / \, B(a+x, b+m-x)$$

$$= B(a+x+k, b+m-x) / B(a+x, b+m-x) .$$

Equation (8.28) now follows using standard results on beta and gamma functions, equations (7.27) and (7.23).

(3) is immediate from (8.28) with $k=1$, and is just the mean of the $Be(a+x, b+m-x)$ distribution.

(4) The beta-binomial distribution (8.30) is another application of the beta-binomial theorem, equation (7.73), using (8.27).

............}

Equations (8.27) and (8.29) are particularly simple, and this general structure of exchangeable propositions with a beta prior distribution is extremely powerful.

{............

Example

A doctor proposes a new treatment protocol for a certain form of cancer. With the current treatment approximately 40% of patients survive to six months after diagnosis. The doctor believes that for the modified treatment the success rate will be slightly higher. Let $S_i \equiv$ 'i-th patient receiving new treatment protocol survives six months'. The doctor judges the S_is to be exchangeable given her current information H. (This is a natural judgement but we should note that she might also have allowed for some time being required to learn to use the new treatment effectively, and $P(S_2|H)$ would then be slightly higher than $P(S_1|H)$. The difference would be small in any case, and we ignore this detail.) Then T is the future success rate of the new treatment, and she measures

$$E_1(T|H) = 0.45 . \tag{8.31}$$

She believes that T will probably lie between 0.35 and 0.57 and so assigns a measured standard deviation of 0.11:

$$var_1(T|H) = 0.012 . \tag{8.32}$$

The unimodal beta shape is appropriate, and therefore she asserts

$$T|H ~ \sim ~ Be(a,b) ,$$

with a and b chosen to solve

$$\frac{a}{a+b} = 0.45 , \qquad \frac{ab}{(a+b)^2(a+b+1)} = 0.012 ,$$

$$\therefore a+b+1 = 0.45 (1-0.45)/0.012 = 20.625 ,$$

$$\therefore a = 0.45 (20.625 - 1) = 8.83 ,$$

$$\therefore b = 10.8 .$$

Allowing for the error in her original measurements (8.31) and (8.32), she rounds these values and finally states

$$T|H \;_2\sim\; Be(9,11) . \tag{8.33}$$

Her implied prior mean is still 0.45 and the variance is 0.012 correct to three decimal places.

A year later, fifteen patients have received this treatment and have been followed through to six months after diagnosis. Of these fifteen, six have survived. From (8.27) the doctor's belief about T is now

$$T|(X=6)\wedge H \;_2\sim\; Be(15,20) .$$

Her expectation of T, or equivalently her probability that any subsequent patient under the new treatment will survive six months, is

$$P_2(S_i|(X=6)\wedge H) = 15/35 = 0.429 ,$$

for $i=16,17,\ldots$ as in equation (8.29). Her confidence in the new treatment has diminished somewhat.

$$var_2(T|(X=6)\wedge H) = 15\times 20/(35^2\times 36) = 0.0068 ,$$

corresponding to a standard deviation of 0.082. Therefore she has a probability of approximately two-thirds that T lies within 0.082 of 0.429, i.e. in the interval $[0.347, 0.511]$. Compare this with the corresponding prior interval of $[0.35, 0.57]$. She no longer gives much probability to $T > 0.5$, and still considers it a distinct possibility that this treatment protocol is worse than the old one ($T < 0.4$). The results are not encouraging to the doctor, but substantially more data are required before this treatment could be said to be properly evaluated.

..........}

Provided a and b are sufficiently large – both at least ten, say – the distribution of $Be(a,b)$ is close to that of the normal random variable having the same mean and variance (see Section 7.9). We could therefore use tables of the standard normal distribution to approximate posterior probabilities for T.

Exercises 8(c)

1. Let T be the proportion of broad-leaved trees in a forest. Your initial beliefs about T are expressed in $E_1(T|H)=0.4$, $var_1(T|H)=0.02$. You observe 20 trees and find that 11 are broad-leaved. Summarize your updated beliefs about T.

2. Consider a long series of coin tosses divided into groups of 6 tosses. Let S_i be the proposition that the i-th group shows three heads and three tails. Regarding the S_is as exchangeable, measure your prior distribution for T using the beta family. Now refer to Exercises 1(a)4-6, where you performed some of these

tosses, and compare your probabilities (updated through Bayes' theorem and de Finetti's theorem) with the direct measurements you made then.

Now use the binomial probabilities theorem to obtain an objective probability for S_i. How does this calculation relate to your beliefs about T?

3. Let R_i be the proposition of a red ball in the i-th draw from Polya's urn (Exercises 1(c)2, 8(a)4-5, 8(b)1), for $i = 1, 2, 3, \ldots$. Prove that your beliefs about the R_is correspond to them being exchangeable with prior distribution $T|H \sim \text{Be}(1, 1)$. Show that an exchangeable sequence with a $\text{Be}(a, b)$ prior distribution can also be described in terms of Polya's urn.

4. A geologist examines the sizes of pieces of quartz found in a certain rock. Slices are cut through the rock and the areas of individual quartz sections are measured. He is interested in $A \equiv$ the proportion of sections that have areas less than one square centimetre. His prior expectation of A is 0.9, and he would be very surprised if A were less than 0.8. Fifty slices through the rock yielded 215 quartz sections, of which 17 had areas greater than one square centimetre. Using beta distributions, measure his prior and posterior distributions for A. If the next slice has six quartz sections, calculate his posterior probability that they will all have areas less than one square centimetre.

8.6 Probability and frequency

In an infinite exchangeable sequence we have defined T to be the proportion of true propositions in the entire sequence. If You know the value, t, of T then the propositions become trials. They are mutually independent with a common probability t. In this case there is nothing for You to learn about any E_i from observing any other E_js. We have remarked earlier that this inability to learn is unrealistic in practice. The reason is that T is almost always unknown. You can never observe T. However many of the propositions You observe You cannot see them all, and therefore You will not know T.

Nevertheless, if You observe a great many of the E_is it is clear that You will know much more about T than if You observe none of them. If an engineer observes that in 5000 lifetime tests of a particular bearing the lifetimes exceed 1000 hours in 4652 tests, then he will surely believe that the proportion of all such bearings with lifetimes over 1000 hours is close to the observed *frequency* $4652/5000 = 0.9304$. In this section we explore how Your beliefs about T, and about unobserved E_is, become dominated by the observed frequency $p \equiv x/m$, when the number m of observed propositions becomes large.

If You begin with a beta prior distribution for T then this phenomenon is easy to study using the results of Section 8.5. First consider Your posterior expectation of T,

$$E(T|(X=x)\wedge H) = (a+x)/(a+b+m) . \tag{8.34}$$

We suppose that $m \to \infty$ and $x \to \infty$ such that $x/m \to p$. It is obvious that the numerator of (8.34) will be dominated by x and the denominator by m, so that the ratio will tend to p. We can derive this limit formally as follows.

$$E(T|(X=x)\wedge H) = \frac{x}{a+b+m} + \frac{a}{a+b+m} \,. \tag{8.35}$$

The second term in (8.35) tends to zero as $m \to \infty$. Therefore

$$\lim_{m\to\infty} E(T|(X=x)\wedge H) = \lim_{m\to\infty} \{(x/m) \times (1-(a+b)/(a+b+m))\}$$

$$= \lim_{m\to\infty} (x/m) \times \lim_{m\to\infty} (1-(a+b)/(a+b+m))$$

$$= p \times 1 = p \,.$$

So, after a large number of observations Your posterior expectation of T will tend to the observed frequency p. Because of (8.29) Your posterior probability for any unobserved proposition E_i will also tend to p. Furthermore, notice that

$$P(E_{m+2}|E_{m+1}\wedge(X=x)\wedge H) = \frac{P(E_{m+1}\wedge E_{m+2}|(X=x)\wedge H)}{P(E_{m+1}|(X=x)\wedge H)}$$

$$= \frac{q_2^*}{q_1^*} = \frac{a+x+1}{a+b+m+1} \,,$$

which also tends to p as $m \to \infty$. Therefore the unobserved E_is tend towards independence as the number of observations becomes large. We know that they are strictly independent if T is known, and the tendency towards independence here corresponds to the fact that for large m T is 'almost known'. Thus

$$var(T|(X=x)\wedge H) = \frac{(a+x)(b+m-x)}{(a+b+m)^2(a+b+m+1)}$$

$$= \frac{a+x}{a+b+m} \times \frac{b+m-x}{a+b+m} \times \frac{1}{a+b+m+1} \,. \tag{8.36}$$

As $m \to \infty$ the first term of (8.36) tends to p, the second to $1-p$ and the third to zero. The variance therefore tends to zero, and so for large m You have substantial information about T.

All of this theory applies if Your prior distribution for T is a beta distribution. Regardless of the a and b parameters of the prior distribution, the prior information will eventually become dominated by the information in the observations. The same results can be proved to hold for practically any prior distribution. The following is a sketch of how the proof proceeds, omitting technical details. Remember that Bayes' theorem multiplies the prior density by the likelihood function to produce the posterior density. We have the likelihood function

$$P(X=x|(T=t)\wedge H) = \binom{m}{x} t^x (1-t)^{m-x}$$

for $0 \leq t \leq 1$. As m increases, the prior stays constant but the likelihood becomes more and more sharply peaked around $t=x/m$. This is shown in Figure 8.1 for $m=10, 50$ and 250 with $x/m=0.4$. As long as the prior density is not actually zero at and around $t=x/m$ the likelihood will eventually dominate, and the posterior density will also concentrate on $t=x/m$.

The posterior convergence theorem

Let $E_1, E_2, E_3, \ldots$ be exchangeable given H and let T be the proportion of true propositions in the sequence. Let $X_m=x_m$ be the observed number of true propositions in $E_1, E_2, \ldots, E_m$, and let

$$\lim_{m \to \infty} x_m/m = p \ .$$

For some $e > 0$, let T have prior density $d_{T|H}$ satisfying

$$d_{T|H}(t) > 0, \qquad \text{for } p-e < t < p+e \ . \tag{8.37}$$

Then

(1) $\quad \lim_{m \to \infty} E(T|(X_m=x_m) \wedge H) = p \ ,$ $\hfill (8.38)$

(2) $\quad \lim_{m \to \infty} P(\wedge_{i=m+1}^{m+k} E_i | (X_m=x_m) \wedge H) = p^k \ ,$ $\hfill (8.39)$

for $k=1, 2, 3, \ldots,$

(3) $\quad \lim_{m \to \infty} var(T|(X_m=x_m) \wedge H) = 0 \ .$ $\hfill (8.40)$

The condition (8.37) simply requires that You do not believe it completely impossible that T is close to p, on the basis of Your prior information H. Otherwise the data information $X_m=x_m$ would be completely at odds with Your prior information. Under any realistic prior information, and in particular for any beta prior distribution, this condition is satisfied.

The key role of this theorem derives from its very wide applicability. Not only are we often interested in an essentially infinite sequence of related propositions, but also the judgement of exchangeability is typically objective. Different people will only have different beliefs about the E_is because they have different, subjective prior beliefs about T. The posterior convergence theorem says that the more propositions we observe the less effect these subjective differences have on our posterior beliefs. When m is very large then all observers, whatever their prior beliefs, will have essentially the same posterior beliefs.

We now have a new context in which objective probability measurements can be made. That is, in a series $E_1, E_2, E_3, \ldots$ of related propositions, when x out of a large number m of these propositions are known to be true, the measurement of x/m for the probability of any unobserved proposition in the series is

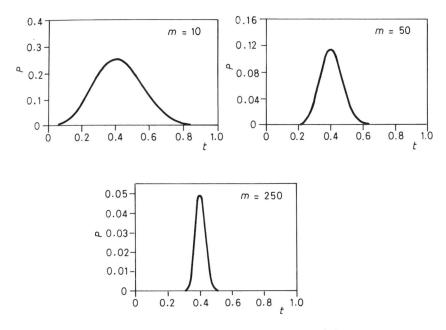

Figure 8.1. Likelihoods for T with $x/m = 0.4$

objective. We use the word 'objective' as usual to mean that nobody would give a noticeably different probability unless they had some special information.

In Section 2.6 we noted that early probability theorists tried to base a definition of probability on equi-probable partitions. The attempt failed because it was far too narrow. It was incapable of assigning probabilities in numerous important contexts. Other mathematicians have attempted to define the probability of a proposition E as equal to the limit p of the frequency of true propositions in a large number of propositions exchangeable with E.

Frequency probability
Let H include the information that in a large number of exchangeable propositions a proportion p are true. Then if E is a further exchangeable proposition

$$P(E|H) = p \ .$$

As a definition of probability, this is also extremely restricted. It can only be used when very substantial frequency information is already available. Inevitably, it would seem, this definition is also doomed. Yet in fact it is still espoused by many statisticians who endeavour to make probability an objective science. Their approach is radically different from the view adopted in this book. It asserts that probabilities may exist (because an appropriate infinite exchangeable sequence exists) but are unknown (because the sequence has not been observed). The resulting convolutions are complex. A probability statement in this approach does not apply to any proposition of interest, but only has validity as one of an infinite series of such statements made in a series of propositions exchangeable with the one of interest.

The personal probability definition adopted here agrees with both symmetry probability and frequency probability whenever they are applicable. Otherwise an objective probability judgement is not possible.

Another consequence of the posterior convergence theorem is that when You have substantial frequency data available You need not think carefully about Your prior distribution for T. Strictly, as $m \to \infty$ all prior distributions will yield identical posterior distributions. For large but finite m some prior beliefs will still be sufficiently strong for the prior information not to be dominated by the observations. For instance, with a beta prior distribution, this will happen if $a+b$ is not very much less than m. In practice, then, provided Your prior information is weak relative to the observed frequency data then any relatively weak prior distribution may be chosen without significantly affecting the accuracy of Your posterior distribution. A convenient prior distribution in this case is

$$T|H_1 \sim Be(1,1) \sim Uc(0,1).$$

Then

$$T|(X=x) \wedge H_1 \sim Be(x+1, n-x+1)$$

and

$$E_1(T|(X=x) \wedge H) = \frac{x+1}{m+2} \approx \frac{x}{m}, \tag{8.41}$$

$$var_1(T|(X=x) \wedge H) = \frac{(x+1)(m-x+1)}{(m+2)^2(m+3)}$$

$$\approx \frac{1}{m} \times \frac{x}{m}\left(1-\frac{x}{m}\right). \tag{8.42}$$

The approximation (8.41) is implicit in the earlier discussion in this section. Equation (8.42) is another useful approximation. It may be further simplified by noting that

$$\frac{x}{m}\left(1-\frac{x}{m}\right) \le \frac{1}{4}$$

for any x (between 0 and m). The expression is equal to $1/4$ when x/m equals one-half, and is close to $1/4$ provided x/m is not too close to 0 or m. Therefore

$$var_1(T|(X=x) \wedge H) \approx 1/(4m)$$

will often be a good approximation. it has the advantage of being *conservative*, meaning that it always exceeds Your actual posterior variance, and so does not imply more information than You really have.

Strong frequency information

Let $E_1, E_2, E_3, \ldots$ be an exchangeable sequence given H, and let T be the proportion of true propositions in the sequence. Let X be the observed number of true propositions in a substantial number m of E_is. Then provided Your prior information H is weak relative to the observed data $X=x$, the following are accurate posterior measurements.

$$E_1(T|(X=x)\wedge H) = x/m ,$$

$$var_1(T|(X=x)\wedge H) = x\,(m-x)/m^3$$

$$\leq 1/(4m) . \tag{8.43}$$

The bound (8.43) provides the following additional measurements.

$$P_1\left(\frac{x}{m} - \frac{1}{2m^{1/2}} < T < \frac{x}{m} + \frac{1}{2m^{1/2}} \,\middle|\, (X=x)\wedge H\right) \geq 0.65 .$$

$$P_1\left(\frac{x}{m} - \frac{1}{m^{1/2}} < T < \frac{x}{m} + \frac{1}{m^{1/2}} \,\middle|\, (X=x)\wedge H\right) \geq 0.95 .$$

8.7 Calibration

In the United States of America, weather forecasters on television and radio often announce their personal probabilities for the proposition that it will rain the next day. Suppose that a forecaster has made such statements every day over a long period of time. Consider all those days for which he asserted a probability 0.5 of rain, and suppose that on a proportion 0.3, i.e. 30%, of those days it rained. The discrepancy between the stated probability 0.5 and the observed frequency 0.3 suggests, on a purely intuitive level, that he is not forecasting well. The intuition is supported by the frequency arguments of Section 8.6.

Suppose that You observe the weather forecaster over this period of time. Assuming that You have no special knowledge of meteorology or local weather conditions, it is reasonable for You to regard all those days on which the forecaster gave his probability as 0.5 to be exchangeable. After observing a relative frequency of 0.3 in a large number of such propositions, the next time the forecaster says that his probability is 0.5 for rain, Your probability will be 0.3. In general, if the forecaster's stated probabilities do not accord with the observed frequencies, then You will not accept his probabilities.

Calibration

You are said to be calibrated if, for every p in $[0, 1]$, the proportion of true propositions among all those to which You give probability p is p.

The weather forecaster's forecasts in the example are not calibrated for $p = 0.5$. This may be because he is a bad meteorologist or a poor measurer of probabilities. It is clearly good to be calibrated. If You hear a forecaster give a probability 0.5 of rain, and You have not observed his past forecasts and have no knowledge of his performance as a forecaster, You do not know how to interpret his forecast. Your probability for rain need not be 0.5. If, however, You can assume that a weather forecaster will be calibrated, then You can adopt whatever probability he asserts as Your own probability. (We continue to assume that You have no special information.)

An uncalibrated forecaster is bad unless You have a long, detailed history of the success of his forecasts from which to interpret his 'probabilities'. A calibrated forecaster is not necessarily good. If the overall proportion of days on which it rains in Your district is 0.4, then a forecaster who gave a probability of 0.4 every day would be calibrated, but he would not be very informative. A good forecaster is calibrated and also gives probabilities as close as possible to zero or one.

One way in which you can improve your own probability measurement skills is to check whether your measurements are calibrated. It is for this reason that regular probability measurement exercises were recommended in Section 1.2. If you find that you are not well calibrated, this suggests that your measurements can be improved. Suppose, for instance, that the proportion of true propositions is observed to be 0.3 among those propositions to which you gave a probability of 0.2. Then, in future, when you feel that your degree of belief in a proposition corresponds to a probability of 0.2, you should perhaps give it the probability 0.3 instead. For it appears that that feeling actually corresponds to the degree of belief which is correctly represented by the probability 0.3.

9

Statistical models

The techniques of Chapter 8 are extended in this chapter to cover the basic principles of statistical modelling. These ideas are discussed first in general terms in Section 9.1. Statistical analysis of data begins by representing the prior joint distribution of all the observations as having some simple structure conditional on certain unknown *parameters*. This is called a statistical *model* and is relatively objective. In contrast, the prior distribution of the parameters is typically subjective. Given sufficient observations, this prior distribution becomes dominated by the data, and the posterior distribution of the parameters is then essentially objective. This is the strength of statistical modelling.

A large class of models begin with a judgement of exchangeability between the data. Sections 9.2 to 9.5 deal with exchangeable random variables generally. They are defined in Section 9.2, and more terminology is introduced here and in Section 9.3. In particular, an infinite sequence of exchangeable random variables is called a *population*, and a set of observations of random variables in a population is called a *sample*. Section 9.2 obtains a version of de Finetti's theorem for exchangeable random variables. The application of de Finetti's theorem is not simple because a very high-dimensional distribution must be measured. Section 9.3 considers the effect of observing a large sample.

Measurement of distributions for random variables can often be improved by thinking of them as contained in a population. This approach is explained in Section 9.4, and in particular expressions for the mean and variance of a random variable are obtained in terms of the mean and variance of the population.

Section 9.5 discusses simplifications of the measurement problem by assuming a *parametric model*. Sections 9.6 and 9.7 analyse two of the simplest parametric models. The normal location model is dealt with in Section 9.6 and the Poisson model in Section 9.7. A different approach is used in Section 9.8. No specific distribution is assumed, but a generally applicable, approximate analysis is developed. Section 9.9 concludes this book by looking very briefly at the two major fields of application for the ideas developed here – Statistics and Decision Theory.

9.1 Parameters and models

The key to our analysis of frequency data in Chapter 8 was conditioning on the unknown quantity, R or T. This separated the subjective, personal judgement of a distribution for R or T from the essentially objective, conditional probabilities of the E_is. This approach is the characteristic feature of *statistical modelling*. Here is another example.

A new fertilizer is developed, and in its first year of testing a standard variety of wheat is grown on 12 different sites. Let X_i be the wheat yield of the i-th site, in tonnes per hectare. A farmer is considering using this fertilizer. He is not directly interested in the X_i s. His concern is with the average yield which he himself will obtain if he adopts this fertilizer and uses it under similar conditions in the future. Denote this long-run average yield by A and the farmer's current information by H. The X_i s are of direct interest because observing them will change his beliefs about A. Ignoring complexities such as the fact that average yields will vary from year to year because of weather variations, the following judgement is essentially objective –

$$E(X_i|H \wedge (A=a)) = a , \tag{9.1}$$

That is, if he knew the long-run average yield then that would be his expected yield for any site. Since this is a standard variety of wheat, the farmer knows how variable its yield will be from site to site. He extends (9.1) by

$$X_i|H \wedge (A=a) \sim N(a, 0.64) , \tag{9.2}$$

for $i=1, 2, \ldots, 12$. Amongst other farmers, whose own information will include the same knowledge about this variety of wheat, (9.2) is also largely objective. The overall variability to which he assigns the variance 0.64, or a standard deviation of 0.8 tonnes per hectare, would be measured similarly by other farmers. They would also agree on the general shape implied by the normal distribution. Finally, once the fertilizer's average effect, represented by A, is known, observing any X_i will not change the farmer's beliefs about any other X_j. Therefore he judges the X_i s to be mutually independent given $H \wedge (A=a)$.

He completes his specification by measuring his distribution for A given H. This will be subjective. Other farmers may have more or less confidence in this fertilizer or in the company which has developed it. Once the data are observed, the farmer derives his (posterior) distribution for A given $H \wedge \wedge_{i=1}^{12} (X_i=x_i)$. This is also subjective, but personal differences will have been greatly reduced by the data, because the distribution of the X_i s given $H \wedge (A=a)$ is objective.

The similarities between this example and the discussion of frequency data in Chapter 8 are clear.

1. There is an unknown quantity (e.g. R or T or A), conditional upon whose value the joint distribution of the data may be measured easily and objectively. This conditional formulation is called a *model*, and the unknown quantity on which the model is conditioned is called a *parameter* of the model.
2. Prior beliefs about the parameter are typically subjective but, through the objectivity of the model, this subjectivity is generally reduced when the data are observed. Given sufficient data, posterior beliefs about the parameter will become objective.

This simple scenario is complicated in practice in various ways. First, there may be more than one parameter. It may be necessary to condition on values for several unknown quantities before measurement of the joint distribution of the data can be made with ease and relative objectivity. Given a prior joint distribution of all the parameters, Bayes' theorem is applied to obtain their joint posterior distribution. Therefore the parameters represent various different aspects of the application that the data will provide information about.

{...........

Example

A second parameter will be required in the above fertilizer example, if variability of yield may also be affected by the fertilizer. Now the X_is will not be independent given $H \wedge (A=a)$, since if the farmer observes $X_1 = x_1$ and x_1 is far from a, then this will suggest an increased variability, which in turn affects his beliefs about $X_2, X_3, \ldots, X_{12}$. He would then introduce a second parameter V such that

$$X_i | H \wedge (A=a) \wedge (V=v) \sim \mathrm{N}(a, v) .$$

It would now be reasonable to suppose that the X_is are independent given $H \wedge (A=a) \wedge (V=v)$.

...........}

A second complication is that it is frequently unrealistic to hope to identify parameters such that the model is fully objective. Particularly in scientific applications there is always the possibility of debate and disagreement, even over the model. In other problems one might simplify the most objective model by assuming that a parameter has a known value. This is reasonable if it will be generally agreed that its value is not far from the assumed value. The advantage is that models with fewer parameters may be analysed to derive appropriate posterior beliefs, more easily than those with many parameters. This benefit may outweigh the loss of objectivity implied by the simplification.

These are just some of the many practical aspects of statistical modelling and analysis that make it a skilled occupation. Our intention in this book is only to give the reader a taste of such matters. For our purposes the following description will suffice.

Statistical models

Given a set of observations which are to be made, a model consists of a relatively objectively measured, joint distribution of the observations conditional on one or more parameters. The parameters are unknown quantities whose joint distribution prior to the observations is relatively subjective.

The remainder of this chapter consists of a study of certain simple models. First, to give more indication of the scope of statistical modelling, we conclude this section with two examples of multi-parameter models.

{...........

Regression modelling

An experiment is performed to see how a braking mechanism is affected by humidity. The mechanism is used repeatedly to stop a heavy rotating drum in an atmosphere with controlled humidity. After a fixed number of braking operations the mechanism is removed and examined. One measurement which is made as part of this examination is loss of weight of the brake pad. At higher humidity engineers expect greater weight loss from the brake pad. The experiment is performed ten times using ten different humidity levels. Let W_i be the weight loss at humidity h_i. If the experiment were repeated at the same humidity h_i we would not expect the weight loss W_i' in the second trial to equal W_i exactly. There will be some fluctuation in wear. So let L_i be the average weight loss in many repetitions at humidity h_i. We begin with 10 parameters, $L_1, L_2, \ldots, L_{10}$ and propose the model

$$W_i | H \wedge \bigwedge_{j=1}^{10} (L_j = l_j) \sim N(l_i, 1), \tag{9.3}$$

for $i = 1, 2, \ldots, 10$ and the W_is are mutually independent conditional on these parameters. (A variance of 1 is assumed, but in practice a realistic value would be chosen or the variance might be another parameter.) This model has 10 parameters, and has therefore not succeeded in reducing subjectivity at all. Instead of a subjective distribution given H for the 10 observable random variables $W_1, W_2, \ldots, W_{10}$, we require a subjective distribution for the 10 parameters. However, we can reduce the number of parameters by making reasonable assumptions about how weight loss increases with humidity. If a steady increase is expected then we might suppose that

$$L_i = A + B h_i . \tag{9.4}$$

The interpretation of (9.4) is that weight loss increases with humidity at an unknown rate B, starting from an unknown level A at zero humidity. Substituting this into the original model (9.3) gives a new model, that the W_is are conditionally independent with distributions

$$W_i | H \wedge (A = a) \wedge (B = b) \sim N(a + b h_i, 1),$$

$i = 1, 2, \ldots, 10$. This model has only two parameters. Other relationships than (9.4) may be proposed, possibly involving more or fewer parameters. Statisticians use models like this frequently to learn about how one variable influences another.

..........}

{...........

Time series modelling

An amateur meteorologist measures the ozone level outside her house each day for a month. Let R_i be the ozone level on day i. The R_is form a series of observations ordered in time. Let S be the long term average ozone level outside her house at this time of year. Then $E(R_i|H \wedge (S=s))=s$, for $i=1,2,\ldots,30$. An important feature is that she expects amounts of ozone to vary slowly from day to day. In particular, if the level is high on day i then it will tend to be high on day $i+1$. However, this stability holds only over short periods, so that a high ozone level on day i is not a strong indicator that it will be high on day $i+7$. These facts can be described using the concepts of covariance and correlation from Section 6.5. The correlation between R_i and R_{i+1} is relatively high and positive, but the correlation between R_i and R_{i+j} decreases as j increases .

There are many ways of modelling these correlation characteristics, but the following is the simplest model that the meteorologist might use. There are three parameters – S represents the long-run average ozone level and V its long-run variability, while C is a parameter between 0 and +1 representing the correlation between adjacent observations.

$$E(R_i|H \wedge (S=s) \wedge (V=v) \wedge (C=c)) = s ,$$

$$var(R_i|H \wedge (S=s) \wedge (V=v) \wedge (C=c)) = v ,$$

$$cov(R_i,R_{i+j}|H \wedge (S=s) \wedge (V=v) \wedge (C=c)) = v\,c^j , \qquad (9.5)$$

for $i=1,2,\ldots,30$ and for $j=1,2,\ldots,30-i$. Since c is less than one, c^j decreases as j increases. Equations (9.5) do not fully describe the model, since we require to measure the joint distribution of the R_is, not just their means, variances and covariances. Although we have not considered them in this book, there exist several families of standard joint distributions which might be used to complete this model.

...........}

9.2 Exchangeable random variables

Consider again the fertilizer example of section 9.1. The farmer might try to formulate his joint distribution for the X_is by using the idea of exchangeability. His beliefs about the X_is have the same kind of symmetry as probabilities for exchangeable propositions. He has the same knowledge about X_1, the wheat yield on the first plot, as about any other X_i. Therefore his distribution for X_1 is the same as for any other X_i. We can write this as

$$X_1|H \sim X_2|H \sim \cdots \sim X_{12}|H . \qquad (9.6)$$

Furthermore, his joint distribution for any pair of X_is is the same as for any other pair, and so on. He can express this symmetry by saying that the X_is are *exchangeable random variables* given his information H.

Exchangeable random variables

The random variables $X_1, X_2, \ldots, X_n$ are said to be exchangeable given H if, for any $m = 1, 2, \ldots, n-1$ and for any m distinct integers $i_1, i_2, \ldots, i_m$,

$$X_{i_1}, X_{i_2}, \ldots, X_{i_m} | H \sim X_1, X_2, \ldots, X_m | H .$$

That is, for all $x_1, x_2, \ldots, x_m$,

$$P((X_{i_1} \leq x_1) \wedge (X_{i_2} \leq x_2) \wedge \ldots \wedge (X_{i_m} \leq x_m) | H)$$

$$= P((X_1 \leq x_1) \wedge (X_2 \leq x_2) \wedge \ldots \wedge (X_m \leq x_m) | H) .$$

An infinite sequence of random variables $X_1, X_2, X_3, \ldots$ are said to be exchangeable given H if $X_1, X_2, \ldots, X_n$ are exchangeable given H for all $n = 2, 3, \ldots$.

The farmer will observe $X_1, X_2, \ldots, X_{12}$ but is interested in them as part of an infinite exchangeable sequence $X_1, X_2, X_3, \ldots$, containing also the wheat yields in all similar applications of this fertilizer in the future. This is typically the case in practice, and we shall deal only with infinite exchangeable sequences in this section. The judgement of exchangeability for random variables is generally objective. Wherever there is a set of similarly defined quantities X_i then anyone who does not have any special information relating to individual X_is or groups of X_is will judge them to be exchangeable.

{...........

Examples

A company makes turbine shafts for aircraft engines. If T_i is the number of flying hours before routine inspection shows that the i-th shaft needs replacing, then the T_is might be judged exchangeable. The following kinds of information, however, would make exchangeability inappropriate.

1. It might be known that aircraft carrying some of the shafts are used mainly on short flights, whereas others are on long-haul work. (Type of work could affect the life of the turbine shaft.) Therefore the T_is might form into two different exchangeable groups, one for each type of work.
2. The quality of shafts leaving the factory could change over time. If the T_is are ordered so that the 'i-th shaft' is the i-th to leave the factory, then the T_is will form a time series with a correlation structure as discussed in the last example of section 9.1. In particular the joint distribution of X_i

and X_{i+1} given H will be different from that of X_i and X_{i+10}, thereby violating the exchangeability judgement.

In the absence of any such counter-information, the T_is could be judged exchangeable.

...........}

The judgement of exchangeable propositions in Chapter 8 implied the model described by de Finetti's theorem – that the E_is were independent given $T = t$ with common probability $P(E_i | H \wedge (T=t)) = t$. We can use those results to show that an infinite sequence of exchangeable random variables also has an implied model. Consider the proposition that $X_i \leq x$. If the X_is are exchangeable then this has the same probability as '$X_j \leq x$' for any j. In fact, defining

$$E_i(x) \equiv \text{'} X_i \leq x \text{'} , \tag{9.7}$$

the propositions $E_1(x), E_2(x), E_3(x), \ldots$ form an infinite exchangeable sequence for each x. The proportion $T(x)$ of true propositions in such a sequence now arises as a natural parameter. For the farmer, $T(x)$ is the long-run frequency with which yields less than or equal to x tonnes per hectare will be achieved. We have

$$P(E_i(x) | H \wedge (T(x)=t)) = t , \tag{9.8}$$

Notice that for $x > x'$, if $X_i \leq x'$ then certainly $X_i \leq x$. Therefore the proportion $T(x)$ must be at least as great as $T(x')$, i.e. $T(x) \geq T(x')$. We can see that $T(x)$, as a function of x, looks like a distribution function – it is always between zero and one, and it increases with x. In fact, (9.8) shows that if $T(x)$ is known for all x then it is the distribution function of each X_i. Thus, writing $T(.)=t(.)$ to mean $T(x)=t(x)$ for all x, we have

$$D_{X_i | H \wedge (T(.)=t(.))}(x) = P\{X_i \leq x | H \wedge (T(.)=t(.))\} = t(x)$$

for all x. At this point it is useful to introduce some terminology.

Sample and population

An infinite sequence of exchangeable random variables is called a population. A set of observations of the values of m random variables in a population is called a sample of size m from that population. The function $T(x)$, where for every x $T(x)$ is the proportion of random variables in the population whose values are less than or equal to x, is called the distribution function of the population.

Now de Finetti's theorem says that

$$P(E_i(x) | H) = E\{T(x) | H\} .$$

For each x we have a parameter $T(x)$ and the exchangeability model is that conditional on $H \wedge (T(.) = t(.))$ the X_is are independent with common distribution function $T(x)$.

De Finetti's theorem for exchangeable random variables

Let $X_1, X_2, X_3, \ldots$ be exchangeable given H. For each x $(-\infty < x < \infty)$ let $T(x)$ be the population distribution function. Let $H_1 \equiv H \wedge (T(.) = t(.))$, where '$T(.) = t(.)$' is the proposition that $T(x) = t(x)$ for all $-\infty < x < \infty$. Then the X_is are independent given H_1, with common distribution function

$$D_{X_i | H_1}(x) = t(x)$$

for all x. Thus for any $m = 1, 2, 3, \ldots$, for any m distinct integers $i_1, i_2, \ldots, i_m$, and any $x_1, x_2, \ldots, x_m$,

$$P(\wedge_{j=1}^{m} (X_{i_j} \le x_j) | H) = E(\prod_{j=1}^{m} T(x_j) | H). \tag{9.9}$$

{..........

Proof

First suppose that the exchangeable sequence is finite, comprising $X_1, X_2, \ldots, X_n$. For $i = 1, 2, \ldots, n$, let $F_{i,1} \equiv E_i(x_1)$, $F_{i,2} \equiv E_i(x_2) \wedge \neg E_i(x_1)$, $F_{i,3} \equiv E_i(x_3) \wedge \neg E_i(x_2)$, $\ldots$, $F_{i,m} \equiv E_i(x_m) \wedge \neg E_i(x_{m-1})$, $F_{i,m+1} \equiv \neg E_i(x_m)$. Then for each i, $F_{i,1}, F_{i,2}, \ldots, F_{i,m+1}$ is a partition. The partitions are exchangeable (in an obvious definition), and in particular are similar. De Finetti's theorem is now proved in a similar way to the approach of Chapter 8. We merely sketch the proof here. Let R_j be the number of true $F_{i,j}$s in the set of n X_is. Given $R_1 = r_1, R_2 = r_2, \ldots, R_{m+1} = r_{m+1}$, it is easy to see that the $F_{i,j}$s form a complete deal of n from $(r_1, r_2, \ldots, r_{m+1})$. The probability of any proposition concerning the $F_{i,j}$s, or concerning the $E_i(x_j)$s, may be elaborated by extending the argument to the R_js. Letting $n \to \infty$ we convert from R_j to the proportion $T_j = \lim R_j / n$. Then $T_1 = T(x_1), T_2 = T(x_2) - T(x_1)$ $, \ldots, T_m = T(x_m) - T(x_{m-1}), T_{m+1} = 1 - T(x_m)$. Equation (9.9) may be obtained through this limiting process.

..........}

The model has infinitely many parameters, namely $T(x)$ for all $-\infty < x < \infty$. This fact has considerable practical implications! Before dealing with practical matters, it is important to realize that in principle this model is just like any other. We have already remarked that in practice exchangeability is an objective measurement, therefore the model is objective. We can therefore regard the X_is as independent and identically distributed conditional on their unknown distribution function $T(x)$. The model must then be supplemented by a prior distribution for the population distribution function, which will be subjective (and whose measurement poses the practical problems). Once a prior distribution has been specified then, at least in principle, we can proceed to update beliefs after observing some X_is. Bayes' theorem would be used to derive a posterior distribution for the $T(x)$s. This could be converted into distributions for unobserved X_is by extending the argument as in (9.9). From a practical perspective the model is apparently of little assistance. In order to use it you must measure your distribution for the $T(x)$s. Not only are there infinitely many $T(x)$s to consider but their distribution is further complicated by the logical condition that for any $x > x'$ we must have $T(x) \geq T(x')$. A variety of techniques can be applied to tackle these problems.

First suppose that the X_is are discrete random variables with possible values $x_1, x_2, \ldots, x_n$. Then Your uncertainty about the population distribution function reduces to uncertainty about the $n-1$ values $T(x_1), T(x_2), \ldots, T(x_{n-1})$. (Note that $T(x_n)=1$.) The exchangeable propositions of Chapter 8 can be thought of as being discrete random variables with $n=2$ possible values – 'true' coded as zero and 'false' coded as one. Then $T(0)$ is the proportion of true propositions in the sequence, denoted by T in chapter 8. The case $n > 2$ is important in applications where X_i is naturally discrete, and particularly when we have exchangeable partitions. For instance, an anthropologist may be interested in the frequencies of different finger print types in a certain tribe (to compare with a neighbouring tribe, perhaps). For each member of the tribe he defines a partition of perhaps nine different types, using the right thumb print. A random variable could be defined on each of these partitions by arbitrarily coding each type with a number. The resulting random variables would be exchangeable, and proportions of finger prints in different types could be derived from the $T(x_i)$s.

A simple analysis of this discrete case arises if the prior distribution is chosen from a family of standard joint distributions called the Dirichlet family. The beta family is simply the Dirichlet family when $n=2$. The general case of continuous X_is can be approached with a sophisticated extension of the Dirichlet family, called the Dirichlet process family. However, both the Dirichlet and the Dirichlet process families are rather limited, and cannot describe the shapes and kinds of association encountered in real beliefs for these models. These are difficult problems to handle effectively, and we shall not discuss them further.

Exercises 9(a)

1. Consider the random variables R_i in the 'Time series modelling' example of Section 9.1. Are the modelling statements (9.5) consistent with the judgement that the R_is are exchangeable given $H \wedge (S=s) \wedge (V=v) \wedge (C=c)$?

2. For each of the following sets of random variables, would it be reasonable to regard them as exchangeable?

(a) The transpiration rates (i.e. rates of loss of water through their leaves) of a number of plants of the same species.

(b) The heights of British members of Parliament.

(c) The depth of mud measured at various points in a harbour.

9.3 Samples

Suppose that You observe a large sample of X_is. Then as in section 8.6, Your prior distribution for the $T(x)$s is unimportant. Let $f(x)$ be the number of X_is observed to be less than or equal to x, divided by m. Thus, $f(x)$ is the observed frequency of true propositions $E_i(x)$ defined in (9.7). Since $T(x)$ is the proportion of true propositions in the whole sequence $E_1(x), E_2(x), E_3(x), \ldots$, Your belief about $T(x)$ tends towards certainty that $T(x)$ equals the observed frequency $f(x)$. For large m Your posterior belief is that $T(x)$ is at least close to $f(x)$. The function $f(x)$ is called the *empirical distribution function*. For large m, it approximates to the distribution function $T(x)$, and therefore it also approximates Your distribution function for any unobserved x_i. However, $f(x)$ is always a step function, with jumps at the observed values of the x_is. The x_is may really be continuous random variables, so that $T(x)$ must be a continuous function. In this case $f(x)$ approximates $T(x)$ in the same way as we described in Section 8.2: as m increases the number of steps increases and the height of each step decreases, so that in the limit $f(x)$ becomes continuous.

Empirical distribution function

The empirical distribution function of a sample, also known as the sample distribution function, is the function $f(x)$ which for every x is the proportion of observations in the sample whose values are less than or equal to x. The empirical distribution function is always a step function, with steps at the observed sample values. Given a large sample from a population with distribution function $T(x)$ then, for almost any prior beliefs about $T(x)$, Your posterior joint distribution of the $T(x)$s will give high probability to $T(x)$ being close to $f(x)$, for every x.

We have not attempted to be precise in the above result. A proper theorem would be stated like the posterior convergence theorem of section 8.6. More formally, for any arbitrarily small positive number ε, the phrase 'almost any prior beliefs' means any prior belief giving non-zero probability to the proposition that $T(x)$ lies in the range $f(x) - \varepsilon$ to $f(x) + \varepsilon$ for every x. Your prior belief should not exclude the possibility that the population distribution function will have precisely the values suggested by the empirical distribution function. Then for sufficiently large sample size, Your posterior probability that $T(x)$ lies between $f(x) - \varepsilon$ and $f(x) + \varepsilon$ for every x will be close to one. Therefore given enough data it is not necessary to attempt the difficult task of measuring Your distribution for $T(x)$. You simply use $f(x)$.

Although the population is fully characterized by its distribution function, interest often focuses on just one or two aspects. Consider again the farmer who is concerned about future wheat yields using the new fertilizer. The most important aspect of the long-run performance of the fertilizer is the long-run average yield. If the average of all the X_is in the population is not sufficiently high he would not be interested in the fertilizer. The next most important function is long run variability, for which he would wish to consider some measure of dispersion of the X_is in the population. Accordingly, we can define quantities called population mean and variance, which we will find to be analogous to the mean and variance of a distribution, as introduced in Section 5.6.

Population and sample mean and variance

Let $X_1, X_2, X_3,,,$ be a population. The population mean is denoted by M_X and defined as

$$M_X \equiv \lim_{n \to \infty} \bar{X}_n , \tag{9.10}$$

where

$$\bar{X}_n \equiv n^{-1} \sum_{i=1}^{n} X_i \tag{9.11}$$

is the sample mean of the sample $X_1, X_2, \ldots, X_n$. The population variance is denoted by V_X and defined as

$$V_X \equiv \lim_{n \to \infty} S_n^2 , \tag{9.12}$$

where

$$S_n^2 \equiv n^{-1} \sum_{i=1}^{n} (X_i - \bar{X}_n)^2 .$$

Equations (9.10) and (9.11) define M_X to be the average of all the X_is in the population. For $\bar{X}_n$ is the average of the first n X_is and M_X takes the limit of this average as $n \to \infty$. Similarly, equation (9.13) defines S_n^2 to be a measure of variability among the first n X_is. It is the average squared distance of those X_is from their sample mean $\bar{X}_n$. V_X is obtained by letting $n \to \infty$, and is therefore the average over the whole population of the squared distance of each X_i from the population mean M_X.

The use of the words 'mean' and 'variance' for $\bar{X}_n$, M_X, S_n^2 and V_X is deliberate. Suppose that You observe $X_1 = x_1, X_2 = x_2, \ldots, X_n = x_n$, a sample of size n. You implicitly observe also that $\bar{X}_n = \bar{x}_n$, where

$$\bar{x}_n = n^{-1} \sum_{i=1}^{n} x_i \ ,$$

and $S_n^2 = s_n^2$, where

$$s_n^2 = n^{-1} \sum_{i=1}^{n} (x_i - \bar{x}_n)^2 \ .$$

You have an empirical distribution function $f(x)$ which has steps of height n^{-1} at each x_i. It is therefore the distribution function of a random variable taking possible values $x_1, x_2, \ldots, x_n$ with equal probabilities n^{-1}. Let Y_n be a random variable defined to have such a distribution. Then $E(Y_n) = \bar{x}_n$ and $var(Y_n) = s_n^2$, which can be obtained directly from equations (5.13) and (5.14). In other words, given a sample we can define a corresponding random variable Y_n whose mean is the sample mean and whose variance is the sample variance. Now as $n \to \infty$ the empirical distribution function $f(x)$ tends to the population distribution function $T(x)$. Therefore, M_X and V_X are the mean and variance of a random variable defined to have distribution function $T(x)$.

Conditional mean and variance in a population
Let $X_1, X_2, X_3, \ldots$ be exchangeable given H, with population distribution function $T(x)$, population mean M_X and population variance V_X. Let $H_1 \equiv H \wedge (T(.) = t(.))$ and let m and v be the mean and variance, respectively, of a random variable with distribution function $t(x)$. Then for $i = 1, 2, 3, \ldots$

$$E(X_i | H \wedge (M_X = m)) = E(X_i | H_1) = m \ , \qquad (9.14)$$

$$var(X_i | H \wedge (V_X = v)) = var(X_i | H_1) = v \ . \qquad (9.15)$$

{...........

Proof
We have established that $E(X_i | H_1) = m$ and $var(X_i | H_1) = v$. The remaining parts of (9.14) and (9.15) follow from the irrelevant information theorem of Section 3.1, since $T(.) = t(.)$ implies both $M_X = m$ and $V_X = v$.
...........}

Now let us return to the context in which You have observed a large sample. Your posterior belief about the unobserved X_is is that they are exchangeable with a population distribution function $T(x)$ which You believe is very close to Your empirical distribution function $f(x)$. It now follows that M_X will be close to Your observed sample mean $\bar{x}$ and that V_X will be close to Your observed variance s^2.

{............

Example

The data below comprise measurements of the percentage of ash in 30 wagon-loads of coal.

16.7	16.2	18.3	14.5	12.0	17.6	14.9	12.6	15.5
17.9	18.2	21.5	17.9	21.2	14.2	14.4	22.9	17.5
19.4	17.4	15.8	14.3	17.1	13.6	13.2	9.3	16.4
20.7	23.8	15.0						

The sample mean is found to be 16.67 and the sample variance 10.39. Figure 9.1 shows the empirical distribution function for these data. A further 170 measurements were then obtained, resulting in a sample of 200. Figure 9.2 shows the empirical distribution function, which is clearly becoming close to a smooth curve.

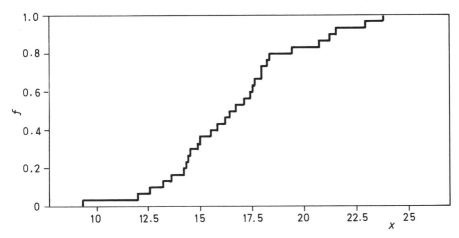

Figure 9.1. Percent ash in coal: first 30 observations

The mean of all 200 observations is 17.04, which we would expect to be closer to the population mean M_X than the mean, 16.67, of the smaller sample. A sample of 200 is not overwhelming evidence, but unless You had strong prior beliefs Your posterior distribution for M_X should be such that You give high probability to values close to 17.04. Similarly, the sample variance in the larger sample is 7.09, and You should believe that V_X is close to this value, rather than 10.39.

............}

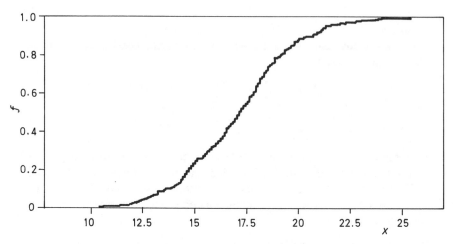

Figure 9.2. Percent ash in coal: 200 observations

9.4 Measuring mean and variance

Equations (9.14) and (9.15), for the conditional means and variances in a population, can be used to obtain unconditional means, variances and covariances.

Mean, variance, covariance of exchangeable random variables
Let $X_1, X_2, X_3, \ldots$ be exchangeable given H, with population mean M_X and population variance V_X. Then, for $i = 1, 2, 3, \ldots$ and $j \neq i$,

$$E(X_i|H) = E(M_X|H), \tag{9.16}$$

$$var(X_1|H) = E(V_X|H) + var(M_X|H), \tag{9.17}$$

$$cov(X_i, X_j|H) = var(M_X|H). \tag{9.18}$$

{..........

Proof
Equation (9.16) follows immediately from (9.14) by using the expectation of conditional expectation theorem of Section 6.6, equation (6.30). Equation (9.17) follows similarly from equations (9.15) and (6.31). The proof of (9.18) is left as an exercise – Exercise 9(b)2 below.
..........}

This theorem is interesting in itself, and we shall return to it in Section 9.8, but it also allows an interesting device for measuring the distribution of any random variable. Suppose that You wish to measure Your distribution for a random variable X, based on Your current information H. Now imagine that X is a member of an infinite population, i.e. You consider a sequence X, X_1, $X_2, X_3, \ldots$ which are exchangeable given H. This is generally not difficult to do. For example, suppose that I have just bought a television, and the manufacturer offers a five year extended warranty for an extra charge. In deciding whether to take this option, I will consider the random variable X = time to first failure of my television. I certainly do not want the extended warranty if X exceeds five years. Thinking about my probability $P(X \leq 5|H)$ it is natural to ask what proportion of televisions of this brand and model will fail within five years. I am then implicitly considering X as contained within the population of failure times for similar televisions.

Your distribution for X given H comprises probabilities $P(X \leq x|H)$. Placing X in the population $X, X_1, X_2, X_3, \ldots$ gives

$$P(X \leq x|H) = E(T(x)|H). \tag{9.19}$$

Therefore, You can formally measure this probability by estimating $T(x)$, the proportion of X_is in the population with values less than or equal to x. The notion of probability is rather abstract, and it can be helpful to think of it in terms of the more concrete idea of a frequency. Thus the probability of a television failing within five years is equated to the frequency of similar televisions failing within five years. Measuring the probability reduces to estimating the frequency. To some extent, this idea is implicit in the direct measurement technique which refers to balls in a bag. There, the probability is equated to the proportion of red balls in a bag which is judged 'equivalent' to the proposition of interest. The step of replacing the balls by propositions exchangeable with the one of interest is both logical and useful. The frequency to be estimated is now more obviously relevant and more concrete.

However, we have remarked in Section 5.7 that it is impractical and inaccurate to measure a distribution by measuring all its component probabilities separately. Instead we measure appropriate summaries, like shape, mean and variance, and then fit a standard distribution. Mean and variance are even more abstract quantities than a probability, and the device of embedding X in a population can also be effective in these measurements. From (9.16) we have

$$E(X|H) = E(M_X|H). \tag{9.20}$$

You may think of the mean of X by equating it to the average value in the population, M_X. The definition of $E(X|H)$ as a weighted average of possible values of X, equation (5.13), is replaced by the simple average M_X. This is a useful analogy even though the population is hypothetical, the mean M_X is not known, and therefore the strict interpretation of (9.20) is to replace one expectation by another. It is helpful to understand expectation in this way – that a typical value of X is also a typical value of the population average M_X.

Moving to the measurement of $var(X|H)$, equation (9.17) is more complex. The fact that it identifies $var(X|H)$ as the sum of two terms is actually its strength. In practice, people tend to underestimate their uncertainty. This phenomenon can be alleviated by measuring separately the two components of variance on the right-hand side of (9.17). First, You need to estimate how much variability would be seen amongst the members of the population. Estimating this variability gives a measurement of $E(V_X|H)$. Next, You must assess Your uncertainty about the population average, yielding a measurement of $var(M_X|H)$. Remember that in both cases we are dealing with squared distances. Thus $E(V_X|H)$ is a typical squared distance of a population member from the mean M_X. Likewise, measuring $var(M_X|H)$ requires measuring $sd(M_X|H)$ by the 'rule of thumb' technique of section 5.7, and then squaring it.

Measurement by exchangeability

$P(E|H)$ may be measured by contemplating a sequence of propositions $E, E_1, E_2, E_3, \ldots$, exchangeable given H. Measuring $P(E|H)$ then consists of estimating the proportion of true propositions in the sequence. This device could be used to measure the distribution function $P(X \le x|H)$ of a random variable X, but in practice it is better to measure summaries of the distribution. Then X is imagined as part of a sequence of random variables $X, X_1, X_2, X_3, \ldots$ exchangeable given H, with population mean M_X and population variance V_X. Measuring $E(X|H)$ then consists of estimating the population average M_X. A measurement of $var(X|H)$ is the sum of two parts. The first part is an estimate of the population variability V_X. The second is a measurement of $var(M_X|H)$.

The two components of $var(X|H)$ can be seen also by imagining the population being generated in two stages. First an average value M_X is chosen. Then the X_is are generated by adding random disturbances above and below M_X. These disturbances have average value zero, so that the population mean is not affected, and average squared value V_X. Then, Your uncertainty about an individual value X in the population is composed of uncertainty about what mean value M_X was chosen, plus uncertainty about what disturbance from M_X was given to create this value.

{ ....

Example

A company proposes to market a new product. The key random variable is $S \equiv$ number of sales in the first three months. The marketing manager is required to measure his distribution for S, based on his knowledge and experience H. We suppose that he can imagine a population $S, S_1, S_2, S_3, \ldots$

of exchangeable random variables. The S_is might be first quarter sales of similar products when launched in a similar market. (The difficulty of unique marketing decisions is partly explained by the difficulty of imagining exchangeable random variables.) He estimates that the average of the S_is, i.e. the average first-quarter sales in similar launches, would be 3000. Thus, $E_1(S|H)=3000$. However, this average could be lower than 1500 or higher than 4500. He decides on $sd_1(M_X|H)=1200$, therefore $var_1(M_X|H)=1440000$. In addition, launches of similar products can produce quite variable sales, and he expects that individual S_is might vary by 500 on either side of M_X. Hence he measures $E_1(V_X|H)=500^2=250000$. These measurements provide $var_1(S|H)=1690000$, therefore $sd_1(S|H)=1300$.
..........}

Exercises 9(b)

1. The following data are of length of forearm (in inches) of thirty adult males.

17.3	18.4	20.9	16.8	18.7	20.5
17.9	20.4	18.3	20.5	19.0	17.5
18.1	17.1	18.8	20.0	19.1	19.1
17.9	18.3	18.2	18.9	19.4	18.9
19.4	20.8	17.3	18.5	18.3	19.4

Draw the empirical distribution function for this sample. Calculate the sample mean and sample variance for (a) the first ten observations, and (b) the whole sample.

2. Prove equation (9.18) using the theorem you proved in Exercise 6(d)3.

3. Let X and Y be contained in a sequence of random variables which are exchangeable given H. Show that if the sequence is infinite then $corr(X,Y|H) \geq 0$. Is this result also true if the sequence is finite?

4. Let T be the time in seconds for the winner of the next Olympic 800 metres final. By considering T as one of an infinite exchangeable sequence, measure your mean and variance for T.

9.5 Exchangeable parametric models

We return now to the problem of specifying a prior distribution for the population distribution function $T(x)$ of a sequence of exchangeable random variables. We saw at the end of section 9.2 that if the random variables are discrete with possible values $x_1, x_2, \ldots, x_n$ the problem simplifies to specifying a prior joint distribution for $T(x_1), T(x_2), \ldots, T(x_n)$. In this case a convenient family of standard distributions can sometimes be used.

In Section 9.3 we found that if a sufficiently large sample has been observed then uncertainty about $T(x)$ is negligible. It is then reasonable to treat unobserved X_is as essentially independent with a known distribution function. In particular, uncertainty about the population mean and variance, M_X and V_X, is also negligible, and they may be equated with the sample mean and variance. When neither of these simplifying circumstances holds, it is an impossibly large task to specify a prior distribution for the infinitely many random variables $T(x)$. It is necessary to reduce the problem so that measurement of such a distribution can be implied by making only a small number of actual measurements. The answer may be found by referring back to our opening discussion in this chapter, in Section 9.1. There, an example was presented in which a farmer created a model for experimental wheat yields $X_1, X_2, \ldots, X_{12}$. From the general perspective of section 9.2 he would judge the X_is to be part of an infinite exchangeable sequence. De Finetti's theorem would then model the X_is as independent conditionally on their common distribution function $T(x)$, and a prior distribution for $T(x)$ would be required. However, in Section 9.1 we adopted a direct approach. The farmer modelled the X_is as conditionally independent with a common distribution, but asserted that the distribution was normal with variance 0.64. He then required a prior distribution for only one parameter, the mean of the normal distribution. This is precisely the kind of reduction of the problem that is needed.

The farmer's model is equivalent to saying that the X_is are exchangeable, but with a special kind of prior distribution for $T(X)$. He is asserting that $T(x)$ is the distribution function of a normal random variable with variance 0.64. He is claiming to be certain that the population distribution function is of that form. The only uncertainty he has about $T(x)$ is that he does not know its mean. That mean was denoted by A in section 9.1, but it would now be better to denote it by the symbol M_X that we have adopted generally for the population mean.

Clearly this amounts to very strong prior knowledge about $T(x)$. If taken literally, it is unreasonable strong, for he should allow the possibility that $T(x)$ is not of normal form, or at least that V_X is not equal to 0.64. Nevertheless it is a justifiable approximation. His experience does suggest to him that the population of yields will be normal in shape and that the variability in the population will be represented by a population variance 0.64. He may not be absolutely certain of either suggestion but they are reasonable measurements and lead to some kind of measured distribution for $T(x)$.

The farmer's assumption of a normal distribution for $T(x)$ is an example of the most common way of defining a prior distribution for a population distribution function. The general idea is to assume that $T(x)$ is the distribution function of a random variable from some family of standard distributions. Uncertainty about $T(x)$ then reduces to uncertainty about the parameters of the family. These become the model parameters, where we now use the word 'parameter' in the sense of Section 9.1. Sometimes, as in the farmer's statement that the variance is 0.64, the number of model parameters is further reduced by assigning 'known' values to one or more parameters of the family.

Parametric models for exchangeable random variables

Let $X_1, X_2, X_3, \ldots$ be exchangeable given H, with population distribution function $T(x)$. A parametric model for the population asserts that, conditional on parameters $P_1, P_2, \ldots, P_k$ the X_is are mutually independent and that

$$X_i | H \wedge \wedge_{j=1}^{k} (P_j = p_j) \sim Q(p_1, p_2, \ldots, p_k), \qquad (9.21)$$

where the random variable $Q(p_1, p_2, \ldots, p_k)$ is defined to have a distribution from some family of standard distributions. The model parameters $P_1, P_2, \ldots, P_k$ are (some or all of) the parameters defining that family.

The parametric model amounts to making the subjective assumption that $T(x)$ is certainly a member of the stated family.

The farmer's model (9.2) obviously corresponds to (9.21). It is a parametric model assuming the normal family, and further assuming a known variance. This model is explored more fully in Section 9.6.

A parametric model for a population in which there is only one parameter, and where that parameter equals the population mean M_X, is called a *location* model. Equation (9.2) is an example of a location model. Location models are the simplest of all models which are of real interest in statistical applications. They provide a useful introduction to statistical methods. In addition to the normal location model examined in Section 9.6, we consider the Poisson location model in Section 9.7. In both these models the only unknown parameter is M_X, so naturally the objective is to make posterior statements about M_X based on observing a sample. In Section 9.8 we consider what can be said about M_X when a parametric model is not assumed. The assumption of a parametric model is typically subjective. It cannot be known that $T(x)$ will definitely be a member of some family. Therefore any analysis based on such an assumption can only be an approximation. The benefit of parametric models is that they are enormously more practical than attempting to define a realistic prior distribution for $T(x)$. Furthermore, the model may genuinely be justifiable as an assumption, because Your belief may be that $T(x)$ is almost certain to lie close to the distribution function of some member of the assumed family. Such a belief may even be essentially objective, as in the farmer's model where the normal distribution and assumed variance were based on experience with wheat yields.

It is easy to justify a parametric model if the assumed family has enough parameters and a sufficient variety of shapes to allow it to approximate to a very wide range of other distributions. In this sense, location models are rarely justifiable because they do not allow the assumed distribution to approximate $T(x)$ well if V_X is not close to the assumed value. It is generally more realistic to use a family, such as the full normal family, which contains members with all possible combinations of mean and variance. Such models are called *location and scale models*.

In any complex problem, if progress is to be made towards a solution it may be necessary to make assumptions. The resulting solution can only be as good as the assumptions on which it is based. Part of the skill involved in statistical modelling consists of formulating reasonable (i.e. as objective as possible) assumptions.

9.6 The normal location model

Let $X_1, X_2, X_3, \ldots$ be exchangeable given H. Assume the *normal location model* that the X_is are mutually independent given $H \wedge (M = m)$ with

$$X_i | H \wedge (M = m) \sim N(m, v), \tag{9.22}$$

for $i = 1, 2, 3, \ldots$ and for all m. The conditional variance v is known. We require Your (prior) distribution for M given H. You might have any prior distribution at all for M, but the simplest analysis results if Your distribution is again a member of the normal family. Therefore we assume

$$M | H \sim N(m_0, w_0) \tag{9.23}$$

for some m_0, w_0, and shall not attempt to deal with any other distributions.

The model (9.22) and prior (9.23) together specify Your joint distribution on the X_is. Now suppose that You observe a sample of size n. Let

$$H_n \equiv H \wedge \bigwedge_{i=1}^{n} (X_i = x_i) \tag{9.24}$$

be Your information after observing the sample. (Because of exchangeability it does not matter which X_is comprise the sample. It is notationally convenient to suppose that they are $X_1, X_2, \ldots, X_n$.) Since the population mean, M, is the one unknown parameter, You will certainly wish to derive Your posterior distribution of M. This can be obtained using the normal-normal theorem of Section 7.9, and supposing that the sample observations arrive one at a time. Taking (9.24) to define H_n for all $n = 1, 2, 3, \ldots$, then $H_1 = H \wedge (X_1 = x_1)$ is Your information after the first observation arrives. From (9.22) and (9.23), applying the normal-normal theorem result (7.70), we have

$$M | H_1 \sim N(m_1, w_1), \tag{9.25}$$

where

$$w_1 = v \, w_0/(v + w_0) \,, \tag{9.26}$$

$$m_1 = (vm_0 + w_0 x_1)/(v + w_0) \,. \tag{9.27}$$

Next You observe X_2, and Your information is now $H_2 = H_1 \wedge (X_2 = x_2)$. Inserting (9.22) and (9.25) into (7.70) we obtain

$$M | H_2 \sim N(m_2, w_2) \,,$$

$$w_2 = v \, w_1/(v + w_1) = v^2 w_0/\{v(v + w_0) + vw_0\}$$

$$= v \, w_0/(v + 2w_0) \,, \tag{9.28}$$

$$m_2 = (vm_1 + w_1 x_2)/(v + w_1) = \frac{vw_0 x_2 + v(vm_0 + w_0 x_1)}{v(v + w_0) + vw_0}$$

$$= \frac{vm_0 + w_0(x_1 + x_2)}{v + 2w_0} \,. \tag{9.29}$$

Clearly, as each new observation arrives we can use (7.70) to update Your beliefs about M. When the whole sample is in, we will have $M | H_n \sim N(m_n, w_n)$, for some m_n and w_n. We could easily guess formulae for m_n and w_n based on (9.26) to (9.29), and verify it by mathematical induction. This would be one way of proving the following theorem. The proof following the theorem is an alternative approach which takes the whole sample together and updates in one step.

Normal location model, normal prior distribution

Let $X_1, X_2, X_3, \ldots$ be exchangeable given H. Let the normal location model (9.22) be assumed and let the distribution of M given H be normal as in (9.23). A sample of size n is observed yielding posterior information H_n defined in (9.24). Then

$$M | H_n \sim N(m_n, w_n) \,, \tag{9.30}$$

where

$$w_n = v \, w_0/(v + nw_0) \,, \tag{9.31}$$

$$m_n = \frac{vm_0 + nw_0 \bar{x}_n}{v + nw_0} \,, \tag{9.32}$$

$$\bar{x}_n = n^{-1} \sum_{i=1}^{n} x_i \,.$$

For any $i = n+1, n+2, \ldots$

$$X_i | H_n \sim N(m_n, v + w_n) \,. \tag{9.33}$$

{............

Proof

Since the X_is are independent given $H \wedge (M=m)$ their joint conditional density is the product of their individual conditional densities, which are normal densities given by (9.22). Thus

$$d_{X_1, X_2, \ldots, X_n | H \wedge M}(x_1, x_2, \ldots, x_n, m) = \prod_{i=1}^{n} d_{X_i | H \wedge M}(x_i, m)$$

$$= \prod_{i=1}^{n} (2\pi v)^{-\frac{1}{2}} \exp\{-(x_i - m)^2 / (2v)\}$$

$$= (2\pi v)^{-n/2} \exp\{-\Sigma_{i=1}^{n}(x_i - m)^2 / (2v)\} . \tag{9.34}$$

From (9.23),

$$d_{M|H}(m) = (2\pi w_0)^{-\frac{1}{2}} \exp\{-(m - m_0)^2 / (2w_0)\} . \tag{9.35}$$

We now apply Bayes' theorem. The form of the theorem which we require has not been given explicitly, but is an obvious generalization of the form (7.47) for two continuous random variables,

$$d_{M|X_1, X_2, \ldots, X_n, H}(m, x_1, x_2, \ldots, x_n)$$

$$= \frac{d_{X_1, X_2, \ldots, X_n | H \wedge M}(x_1, x_2, \ldots, x_n, m) \, d_{M|H}(m)}{\int_{-\infty}^{\infty} d_{X_1, X_2, \ldots, X_n | H \wedge M}(x_1, x_2, \ldots, x_n, y) \, d_{M|H}(y) \, dy} . \tag{9.36}$$

The numerator is the product of (9.34) and (9.35), i.e.

$$(2\pi v)^{-n/2}(2\pi w_0)^{-\frac{1}{2}} \exp(-q/2) , \tag{9.37}$$

where

$$q = v^{-1} \sum_{i=1}^{n} (x_i - m)^2 + w_0^{-1}(m - m_0)^2 .$$

We now collect terms in m^2 and m, and complete the square in m, as follows

$$q = m^2(v^{-1}n + w_0^{-1}) - 2m(v^{-1}\Sigma_{i=1}^{n} x_i + w_0^{-1}m_0)$$

$$+ (v^{-1}\Sigma_{i=1}^{n} x_i^2 + w_0^{-1}m_0^2)$$

$$= w_n^{-1}(m - m_n)^2 + r ,$$

where w_n and m_n are given by (9.31) and (9.32), and

$$r = v^{-1} \sum_{i=1}^{n} x_i^2 + w_0^{-1}m_0^2 - w_n^{-1}m_n^2 .$$

The denominator of (9.36) is the integral of (9.37) with respect to m. It is simply the constant needed to make the numerator integrate to one (as a density function must). Therefore (9.36) is some constant times $\exp(-q/2)$, which is some constant times

$$\exp\{-(m-m_0)^2/(2w_n)\} , \qquad (9.38)$$

since r is a constant. (9.38) is obviously the normal density asserted in (9.30), apart from the necessary constant, which is $(2\pi w_n)^{-\frac{1}{2}}$.

To prove (9.33) we apply the normal-normal theorem, equation (7.69) to (9.22) and (9.30).

............}

The interpretation of these results is very simple. There are two sources of information about M. The prior information suggests that M is somewhere near the prior mean m_0. The sample information suggests that it is somewhere near the sample mean $\bar{x}_n$. The posterior mean (9.32) compromises between these two suggestions. m_n is a weighted average of m_0 and $\bar{x}_n$. This can be seen better by writing

$$m_n = (1-c)m_0 + c\bar{x}_n , \qquad (9.39)$$

where

$$c = n\, w_0/(v + nw_0) = n\, v^{-1}/(w_0^{-1} + nv^{-1}) \qquad (9.40)$$

so that

$$1-c = w_0^{-1}/(w_0^{-1} + nv^{-1}) . \qquad (9.41)$$

The weight attached to each information source is proportional to its strength. Stronger information is related generally to smaller variance. For this reason, the inverse of the variance is often called the *precision*. The weight attached to m_0 in (9.39) is proportional to the prior precision w_0^{-1}. The weight attached to the sample mean is proportional to nv^{-1}. We can think of this as the accumulated precision of the n sample observations, each of which has precision v^{-1}. The idea of accumulating precisions like this is reinforced by writing (9.31) as

$$w_n^{-1} = (v + nw_0)/(vw_0) = w_0^{-1} + nv^{-1} ,$$

showing that posterior precision is the sum of prior precision and the precision of each observation. We also prove formally in Section 9.8 that the precision of the sample mean is n times the precision of each observation.

The more observations there are in the sample, i.e. the larger n is, the more weight is given to the sample, and so the closer m_n will be to $\bar{x}_n$. Furthermore, as n increases w_n decreases. Ultimately, as $n \to \infty$, You will become convinced that M is arbitrarily close to the sample mean, which was the conclusion proved more generally in Section 9.3. However, we now have a proper quantification of Your beliefs about M when the sample size is large but not infinite. For large n we can approximate m_n by $\bar{x}_n$ and w_n by v/n. These same approximations can be achieved by, instead of letting sample information become very strong, letting prior information become very weak. For, as we let prior precision w_0^{-1} go to zero we again find $m_n \to \bar{x}_n$ and $w_n \to v/n$.

Relatively weak prior information

In the normal location model with normal prior, suppose that the prior information is much weaker than sample information. Then the posterior distribution of M may be approximated by

$$M|H_n \sim N(\bar{x}_n, n^{-1}v).\tag{9.42}$$

As $n \to \infty$ Your uncertainty about M becomes negligible. Notice, however, that Your uncertainty about any unobserved X_i does not disappear. Equation (9.33) shows that the distribution of $X_i|H_n$ is approximated for large n by $N(\bar{x}_n, v(1+n^{-1}))$ and its variance tends to v as $n \to \infty$. The conditional variance v, i.e. the population variance V_X, represents inherent variability in a population. Even if the population distribution function is known, each X_i has a variance equal to the population variance. In (9.33) and more generally in (9.17), we see that given any information Your uncertainty about an X_i is composed of (your estimate of) inherent variability plus Your uncertainty about the population distribution function. Even if the latter uncertainty disappears, inherent variability remains.

{...........

Example

In 1798 the scientist Henry Cavendish reported 23 measurements of the Earth's density,

5.36	5.29	5.58	5.65	5.57	5.53	5.62	5.29
5.44	5.34	5.79	5.10	5.27	5.39	5.42	5.47
5.63	5.34	5.46	5.30	5.78	5.68	5.85	

What should he now believe about the Earth's density? Let D be the true density of the Earth, and let the 23 measurements be part of a population $X_1, X_2, X_3, \ldots$ of such measurements. We will first assume that $D = M_X$, i.e. that the average of the population of measurements is the true density. The assumption is therefore that there was no bias in the measurement technique used by Cavendish. We next assume that the population distribution function is normal, which is reasonable in practice for many kinds of physical measurements.

It is probable that Cavendish had a good idea of how accurate his measurements were. For the purposes of this example, we suppose that he asserts that $V_X = 0.1$. This would imply a probability of about 0.95 that any given observation will lie within $2 \times 0.1^{1/2} = 0.63$ of the true density. These assumptions together imply the normal location model

$$X_i|(D=d)\wedge H \sim N(d,0.1).$$

We now require a prior distribution for D. We cannot know what Cavendish's beliefs were about D before he collected these data, but we will suppose that his information was not strong. We let

$$D|H \sim N(6,4),$$

suggesting that with probability about 0.65 D lies between 4 and 8, and with probability 0.95 it lies between 2 and 10.

The sample mean is $\bar{x}_n=5.485$. Applying the theorem we find

$$D|H_n \sim N(m_n, w_n),$$

where

$$w_n = 0.1 \times 4 / (0.1 + 23 \times 4) = 0.00434,$$

$$m_n = (0.1 \times 6 + 23 \times 4 \times 5.485)/(0.1 + 23 \times 4) = 5.486.$$

This is an example in which even 23 observations prove to be far stronger than the prior information. We see that the posterior distribution is very close to the distribution, (9.42), which results from weak prior information – $E(D|H_n)$ is very close to $\bar{x}_n$ and $var(D|H_n)$ is very close to 0.1/23 $= 0.00435$.

Figure 9.3 compares the prior and posterior distributions. The solid line is the posterior density, showing that after observing the data there is negligible probability that D is as low as 5 or as high as 6. The prior density, shown dotted, is more spread out. The greater strength of the posterior information is shown by the fact that the posterior density is much narrower, and therefore reaches higher (since the area under each curve is unity).

It is interesting also to see how Cavendish's beliefs about D might have evolved as he made each successive measurement. The solid line in Figure 9.4 plots the posterior mean $E(D|H_n)$ for $n=1,2,\ldots,23$. After the first observation, 5.34, his posterior mean is $E(D|H_1)=5.376$, and for the first few observations it changes substantially with each new piece of information. Soon, however, it settles down to between 5.44 and 5.52. The dotted lines in Figure 9.4 show bands of width $2sd(D|H_n)$ on either side of the mean. Therefore, on information H_n he is almost sure that D lies within these bounds. As n increases the bands grow steadily narrower.

...........}

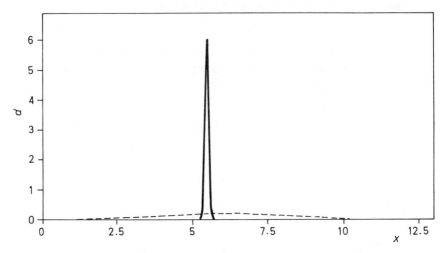

Figure 9.3. Prior and posterior densities for the Cavendish data

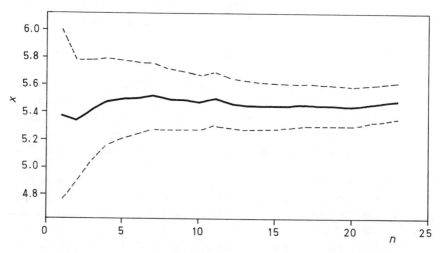

Figure 9.4. Posterior mean against sample size: Cavendish data

Exercises 9(c)

1. A physicist is interested in $L \equiv$ average length of a polymer (in millimetres). He measures his prior distribution as $L|H_1 \sim N(100, 400)$. In an experiment he isolates five polymer strands, straightens and measures them. The measurements are 77, 65, 91, 85, 78, which he assumes are normally distributed given $H \wedge (L=l)$ with mean l and variance 100. Obtain his posterior distribution for L. Can he now claim that $L < 100$?

2. In the normal location model (9.22) with prior distribution (9.23), obtain $E(X_i|H)$, $var(X_i|H)$ and $corr(X_i,X_j|H)$.

3. You have developed a new drug which you hope will help to control arthritis. In experiments on animals infected with arthritis. the approximate size of a cartilage pad is measured twice, once before treatment and again after three months. In appropriate units, the loss of cartilage for untreated infected animals is on average 3.5. Your experiments provide a sample of 20 observations on animals treated with the new drug. You model these as normally distributed given $H \wedge (M=m)$ with mean m and variance 1. The sample mean is $\bar{x}=3.2$. Assuming that your prior information is weak relative to the data, derive your posterior distribution for M. How many more observations would you need to make if you wish your posterior standard deviation to be no greater than 0.1?

9.7 The Poisson model

Now suppose that $X_1, X_2, X_3, \ldots$ are exchangeable given H and we assume that conditional on $M=m$ the X_is are mutually independent with

$$X_i|H \wedge (M=m) \sim \text{Po}(m), \tag{9.43}$$

for $i=1,2,3,\ldots$ and for all m. This is the *Poisson model*. Notice that the X_is are discrete but have infinitely many possible values. Again we suppose that Your prior distribution for M is a member of a convenient family. The gamma family turns out to be convenient, and we suppose that

$$M|H \sim \text{Ga}(a,b). \tag{9.44}$$

You obtain a sample of size n from this population, and we define Your posterior information H_n as in (9.24). The resulting posterior analysis is given in the following theorem.

Poisson model with gamma prior distribution.
Let $X_1, X_2, X_3, \ldots$ be exchangeable given H. The Poisson model (9.43) is assumed and let Your prior distribution for M be a member of the gamma family as in (9.44). Let Your posterior information H_n given a sample of size n be defined by (9.24). Then

$$M|H_n \sim \text{Ga}(a+n, b+n\bar{x}_n) \tag{9.45}$$

where $\bar{x}_n = n^{-1}\sum_{i=1}^{n} x_i$ is the sample mean. For any $i=n+1, n+2, \ldots$,

$$X_i|H_n \sim \text{NB}(b+n\bar{x}, (a+n)/(a+n+1)). \tag{9.46}$$

{..........

Proof

Since the X_is are mutually exclusive given $H \wedge (M = m)$,

$$P(\wedge_{i=1}^{n}(X_i = x_i) \mid H \wedge (M = m)) = \prod_{i=1}^{n} P(X_i = x_i \mid H \wedge (M = m))$$

$$= \prod_{i=1}^{n} m^{x_i} e^{-m} / x_i !$$

$$= (\prod_{i=1}^{n} x_i !)^{-1} m^{n \bar{x}_n} e^{-nm} . \tag{9.47}$$

We apply Bayes' theorem in a form appropriate for n discrete random variables and one continuous one, an obvious generalization of (7.65).

$$d_{M \mid H, X_1, X_2, \ldots, X_n}(m, x_1, x_2, \ldots, x_n)$$

$$= \frac{d_{M \mid H}(m) P(\wedge_{i=1}^{n}(X_i = x_i) \mid H \wedge (M = m))}{\int_{-\infty}^{\infty} d_{M \mid H}(y) P(\wedge_{i=1}^{n}(X_i = x_i) \mid H \wedge (M = y)) \, dy} . \tag{9.48}$$

The numerator, using (9.47) and (9.44), is

$$a^b m^{b-1} e^{-am} \{\Gamma(b)\}^{-1} \{\prod_{i=1}^{n} x_i !\}^{-1} m^{n \bar{x}_n} e^{-nm}$$

$$= a^b \{\Gamma(b) \prod_{i=1}^{n} x_i !\}^{-1} m^{b + n \bar{x}_n - 1} e^{-(a+n)m} . \tag{9.49}$$

The denominator is a constant, as are the first two terms in (9.49). Looking at the last two terms it is clear that this is the density of $\text{Ga}(a+n, b+n\bar{x}_n)$ as stated in (9.45). Equation (9.46) then follows from the Poisson-gamma theorem, (7.75), allowing for the extension of the negative binomial family as described there.

..........}

Equation (9.45) could also have been proved by mathematical induction using the Poisson-gamma theorem, equation (76.76). We now have

$$E(M \mid H_n) = (b + n\bar{x}_n) / (a + n) = (1 - c) m_0 + c \bar{x}_n , \tag{9.50}$$

where $m_0 = b/a$ is the prior mean, $c = n/(a+n)$ and $1 - c = a/(a+n)$. We can again express the weights in terms of precisions, representing amounts of information, but the approach is rather more oblique than for the normal location model. Prior precision is simple enough,

$$\{var(M \mid H)\}^{-1} = a^2 / b .$$

We should next define the precision of a single observation to be the inverse of the population variance, in this case M^{-1} since the Poisson distribution has variance equal to its mean. This quantity is unknown, so instead we take an expectation and define the *mean precision* of a single observation to be

$$\{E(V_X|H)\}^{-1} = \{E(M|H)\}^{-1} = a/b \ .$$

The mean precision of the sample of n observations is therefore na/b. We now see that the weights c and $1-c$ in (9.50) are proportional to mean sample precision and prior precision respectively.

Pursuing this analogy with the normal location model, we would now like to find that posterior precision equals the sum of prior precision and mean sample precision. However, the posterior variance is

$$var(M|H_n) = (b + n \, \bar{x}_n)/(a+n)^2 ,$$

which depends on $\bar{x}_n$. Again we avoid the question of variable precision by taking an expectation. Mean posterior precision is defined to be the inverse of

$$E\{var(M|H_n)|H\} = \{b + n \, E(\bar{X}_n|H)\}/(a+n)^2$$

The required expectation is found in the next section to be equal to $E(M|H) = b/a$. Substituting into the above, we find the mean posterior precision equals the sum of prior precision and mean sample precision as required.

We see the expected behaviour of the posterior distribution in large samples. As $n \to \infty$ the posterior mean tends to $\bar{x}_n$ and the posterior variance to zero. The distribution for an unobserved X_i, equation (9.46), gives

$$E(X_i|H_n) = (b + n \, \bar{x}_n)/(a+n) ,$$

$$var(X_i|H_n) = (b + n \, \bar{x}_n)(a+n+1)/(a+n)^2 .$$

As $n \to \infty$, both tend to $\bar{x}_n$, and indeed the distribution of $X_i|H_n$ will be found to tend to $Po(\bar{x}_n)$. There is intrinsic variability represented by a mean precision $\{E(\bar{X}_n|H)\}^{-1} = a/b$.

When the data are much stronger than the prior information, an approximate posterior analysis is derived by letting the prior precision tend to zero, i.e. $a \to 0$. Then we also require $b \to 0$, so that the prior mean and variance will not become infinite. The result is that $M|H_n$ is approximately distributed as $Ga(n, n\bar{x}_n)$, with approximate expectation $\bar{x}_n$ and variance $n^{-1}\bar{x}_n$.

{ ....

Example

In viewing a section through the pancreas, doctors see what are called 'islands'. The following data are the numbers of islands in 900 pancreatic sections. Thus, no islands were found in 327 sections, one island in 340 sections, and so on.

Islands	0	1	2	3	4	5	6
Frequency	327	340	160	53	16	3	1

We assume a Poisson model, that the numbers of islands in the sections have independent Poisson distributions conditional on their (unknown) mean M.

The 900 observations obviously comprise very strong information, relative to which it is safe to assume that Your prior information is weak. We therefore obtain the posterior distribution

$$M \mid H_n \ \sim \ \text{Ga}(n, n\bar{x}_n),$$

where $n = 900$ and we find $\bar{x}_n = 1.004$. Your posterior mean and variance are

$$E(M \mid H_n) = \bar{x}_n = 1.004,$$

$$var(M \mid H_n) = \bar{x}_n / n = 0.0011.$$

You have a probability of about 0.95 that M lies within $2 \times 0.0011^{\frac{1}{2}} = 0.066$ of 1.004, i.e. in the range 0.938 to 1.070.
..........}

The Poisson model (9.43) is a location model since m is the mean of Po(m). Since m is also its variance, the population variance is also unknown, but we would not regard this as a location and scale model because it does not have separate parameters for mean and variance. The fact that $E(M \mid H_n) \to \bar{x}_n$ as $n \to \infty$ shows that the primary role of m is as the population mean.

9.8 Linear Estimation

In this section we consider how some limited statements can be made about the population mean M_X when only a partial model is assumed. The following judgements form the 'partial model'.

1. $X_1, X_2, X_3, \ldots$ are exchangeable given H.
2. $E(V_X \mid H) = v$.
3. $E(M_X \mid H) = m$.
4. $var(M_X \mid H) = w$.

Statements 2, 3 and 4 refer to Your prior distribution for $T(x)$, but fall far short of specifying the distribution. In particular, this is not a parametric model. $T(x)$ is not assumed to fall into any particular family. Since the prior distribution is not specified, it is not possible to derive a posterior distribution for $T(x)$, and therefore we cannot derive a posterior distribution for M_X. Nevertheless we wish to give some kind of posterior estimate of M_X. It is not unreasonable to ask for such a thing because if the sample size is large we know that the sample mean will be a suitable estimate. For moderate sample sizes we would expect to use as an 'estimate' the kind of compromise between sample mean and prior mean represented by (9.39) or (9.50).

First note that the partial model allows statements about the mean and variance of the sample mean.

Mean and variance of the sample mean

Let $X_1, X_2, X_3, \ldots$ be exchangeable given H, with population mean M_X and population variance V_X. Let $\bar{X}_n$ be the sample mean from a sample of size n. Then

$$E(\bar{X}_n | (M_X = m) \wedge H) = m , \tag{9.51}$$

$$E(\bar{X}_n | H) = E(M_X | H) , \tag{9.52}$$

$$var(\bar{X}_n | (V_x = v) \wedge H) = n^{-1} v , \tag{9.53}$$

$$var(\bar{X}_n | H) = n^{-1} E(V_X | H) + var(M_X | H) . \tag{9.54}$$

{............

Proof

Suppose that the sample comprises the first n observations, so that $\bar{X}_n = n^{-1} \Sigma_{i=1}^{n} X_i$. Then, using (6.7) and (6.9) we have

$$E(\bar{X}_n | (M_X = m) \wedge H) = n^{-1} E(\Sigma_{i=1}^{n} X_i | (M_X = m) \wedge H)$$

$$= n^{-1} \sum_{i=1}^{n} E(X_i | (M_X = m) \wedge H)$$

$$= n^{-1} \sum_{i=1}^{n} m = m ,$$

by virtue of (9.14). This proves (9.51); (9.52) follows from the expectation of an expectation theorem (6.30).

Let $H_1 = H \wedge (T(.) = t(.))$ as before. The X_is are independent, and therefore uncorrelated, given H_1. Therefore, from (6.15), (6.16) and (9.15),

$$var(\bar{X}_n | H_1) = n^{-2} var(\Sigma_{i=1}^{n} X_i | H_1)$$

$$= n^{-2} \sum_{i=1}^{n} var(X_i | H_1)$$

$$= n^{-2} \sum_{i=1}^{n} v = n^{-1} v .$$

Equation (9.53) now follows from the irrelevant information theorem, and (9.54) from the conditional variance theorem (6.31).
............}

Equation (9.53) confirms that the precision of the sample mean is nv^{-1}, as asserted in Section 9.6.

Next we must consider what is meant by an 'estimate' of a random variable. There are many possible interpretations and definitions, so that there is a large body of estimation theory, but we shall consider only one approach. By asserting that a random variable X is estimated by d, you incur an *estimation error* of $Y = X - d$. Since X is unknown, Y is unknown, but one would like Y to be small. The effectiveness of a particular estimate d might be judged by its *mean squared error*

$$S(d) \equiv E(Y^2|H) = E\{(X-d)^2|H\}$$

$$= E(X^2 - 2Xd + d^2|H)$$

$$= E(X^2|H) - 2dE(X|H) + d^2$$

$$= var(X|H) + \{d - E(X|H)\}^2 \tag{9.55}$$

(We have used various results from Chapter 6 here.) Clearly the smallest value which $S(d)$ can have is $var(X|H)$, which it achieves if we choose $d = E(X|H)$. Therefore, minimizing mean squared error is a formal justification of our informally talking about the expectation of a random variable as an estimate of it.

Returning to our problem of estimating M_X, Your prior estimate is $E(M_X|H) = m$. After observing a sample, you would now want to estimate M_X by Your posterior mean $E(M_X|H_n)$. But this quantity is not derivable from the partial model. What of our intuitive notion that a weighted average of prior mean and sample mean would be a reasonable estimate? Since we do not have a posterior distribution we cannot say whether such an estimate is the best based on information H_n, nor even what its posterior mean squared error is. The intuitive appropriateness of such an estimate cannot be tested.

Although we cannot obtain the posterior mean squared error we can, by virtue of the expectation of an expectation theorem, obtain Your prior expectation of its posterior mean squared error. In other words, You do not know how good it is for any particular set of observations $x_1, x_2, \ldots, x_n$, but You can compute how good it will be on average. The argument goes as follows. Consider the estimate

$$d(\bar{x}_n) \equiv c_1 m + c_2 \bar{x}_n , \tag{9.56}$$

where c_1 and c_2 are constants which are to be chosen to obtain the best possible estimator of this type. The criterion on which we will judge $d(\bar{x}_n)$ is its average mean squared error

$$e(c_1, c_2) \equiv E[E\{(M_X - d(\bar{X}_n))^2|H \wedge \bar{X}_n\}|H] \tag{9.57}$$

$$= E\{(M_X - d(\bar{X}_n))^2|H\} . \tag{9.58}$$

The following theorem identifies the best choice of c_1 and c_2.

Linear estimator of population mean

Let $X_1, X_2, X_3, \ldots$ be exchangeable given H, with population mean M_X and population variance V_X. Then the linear estimator (9.56) which minimizes the average mean squared error (9.58) is

$$(1-c)\,m + c\,\bar{x}_n \,, \tag{9.59}$$

where

$$m = E(M_X | H) \,,$$

$$c = n\,w / (v + nw) \,, \tag{9.60}$$

$$w = var(M_X | H) \,, \tag{9.61}$$

$$v = E(V_X | H) \,. \tag{9.62}$$

The average mean squared error of this estimator is $w(1-c)$.

{...........

Proof

From (9.57) we have

$$e(c_1, c_2) = E(M_X^2 | H) - 2E\{M_X d(\bar{X}_n) | H\}$$
$$+ E\{(d(\bar{X}_n))^2 | H\} \,. \tag{9.63}$$

Taking each term in (9.63) in turn we have first

$$E(M_X^2 | H) = var(M_X | H) + \{E(M_X | H)\}^2 = w + m^2 \,. \tag{9.64}$$

Next,

$$-2E\{M_X d(\bar{X}_n) | H\} = -2E[M_X E\{d(\bar{X}_n) | H \wedge M_X\} | H]$$
$$= -2E[M_X(c_1 m + c_2 M_X) | H]$$
$$= -2c_1 m^2 - 2c_2(w + m^2) \,, \tag{9.65}$$

using (6.30) and (9.64). Finally,

$$E\{(d(\bar{X}_n))^2 | H\} = var\{d(\bar{X}_n) | H\} + \{E(d(\bar{X}_n) | H)\}^2$$
$$= c_2^2 \, var(\bar{X}_n | H) + \{c_1 m + c_2 E(\bar{X}_n | H)\}^2$$
$$= c_2^2 (n^{-1}v + w) + (c_1 + c_2)^2 m^2 \,, \tag{9.66}$$

using (9.54) and (9.52). Then adding (9.64), (9.65) and (9.66), (9.63) becomes

$$e(c_1, c_2) = m^2\{1 - 2c_1 - 2c_2 + (c_1 + c_2)^2\} + w(1 - 2c_2 + c_2^2) + n^{-1}vc_2^2$$

$$= m^2(1-c_1-c_2)^2+(n^{-1}v+w)(c_2-c)^2+w(1-c),\qquad(9.67)$$

where c is as in (9.60). Examining (9.67) we see that the first term is minimized by making $c_1+c_2=1$, and the second by making $c_2=c$. The result is (9.59) and the minimal value of $e(c_1,c_2)$ is $w(1-c)$.

.............}

The fact that the best estimator has $c_1+c_2=1$ means that it is a weighted average of m and $\bar{x}_n$. The weights, c and $1-c$, are shown by (9.60) to be proportional to the precisions of m and $\bar{x}_n$, as in (9.40) and (9.41). Indeed, the best linear estimator is exactly the same as the posterior mean of M_X in the normal location model with normal prior distribution. Whenever the posterior mean of M_X is a linear combination of prior mean m and sample mean $\bar{x}_n$, it will exactly equal the best linear estimator. This is true of the Poisson model with gamma prior distribution. In Section 9.7 the quantity w in (9.61) is the prior precision a^2/b, and v in (9.62) is the mean precision of a single observation, a/b. Then (9.60) agrees with (9.50).

In other models the posterior mean will not be a linear function like (9.56). Then the best linear estimator will be a less accurate estimator than the posterior mean, in the sense that its mean squared error will be larger. Its justification is as a simplifying approximation. In a wide variety of models the posterior mean will be close to a linear estimator, and therefore the best linear estimator will be almost the best of all estimators. The extra work required to improve the estimator is the work required to complete the specification of the model by measuring Your full prior distribution for $T(x)$. The extra accuracy may not be worth the extra effort (if the only requirement of the analysis is to estimate M_X).

Remembering that in general the accuracy of an estimator d is measured by its mean squared error $S(d)$, we notice that the best estimate $E(X|H)$ has mean squared error $var(X|H)$. Strength of information about a random variable is therefore directly related to the accuracy with which it can be estimated. The actual posterior accuracy of the best linear estimator is not known but its average mean squared error is $w(1-c)$. Since c lies between zero and one, this is less than the prior variance w. The larger the sample, the nearer c is to one, and therefore the greater the average accuracy of the best linear estimator. In the normal location model with normal prior distribution, $w(1-c)$ is also the actual posterior variance for any sample, equation (9.31). This is the only model for which this happens. In the Poisson model with gamma prior distribution the posterior variance depends on $\bar{x}_n$, but its average value is found to equal $w(1-c)$.

Exercises 9(d)

1. A sociologist is studying recidivism. She has a sample of 15 people who were convicted of arson and released from prison 20 years previously. Let $R_i \equiv$ number of subsequent convictions for i-th person ($i = 1, 2, \ldots, 20$). She assumes the Poisson model

$$R_i | H \wedge (A = a) \sim Po(a),$$

and is interested in the population mean recidivism rate A. Her prior expectation and variance are measured as $E_1(A | H) = 1.5$, $var_1(A | H) = 2$. Fit a gamma prior distribution to these measurements. The observed data are

0, 1, 0, 2, 3, 0, 0, 1, 2, 2, 0, 1, 2, 1, 1.

Derive her posterior distribution for A. What weight has been given to her prior information relative to the data? Re-analyse assuming very weak prior information. Compare the two analyses.

2. Let $X_1, X_2, X_3, \ldots$ be exchangeable given H, with population mean M_X and population variance V_X. Let $\bar{X}_n$ be the sample mean from a sample of size n. Using (9.51) show that

$$E(\bar{X}_n M_X | H) = E(M_X^2 | H),$$

and hence that

$$cov(\bar{X}_n, M_X | H) = var(M_X | H).$$

3. The following partial model is proposed for calibration errors in an instrument. If $S_i \equiv$ squared error on i-th test, then the S_is are assumed exchangeable with population mean Q and population variance R. The quantity Q represents the overall inaccuracy of the instrument and R the variation of that inaccuracy over the calibration range. The engineer specifies

$$E_p(Q | H) = 0.2, \quad var_p(Q | H) = 0.008, \quad E_p(R | H) = 0.005.$$

A series of 25 tests yields a sample mean squared calibration error of 0.34. Estimate Q.

9.9 Postscript

Every book must end, and in doing so must leave many things unsaid. Further development of the ideas and results in this book will take the reader into the fields of Statistics and Decision Theory.

Chapters 8 and 9 have comprised a basic introduction to the theory and methods of Statistics. It is important to realize that there is an ongoing debate about the foundations of Statistics. Since the beginning of this century the major school of thought has based its theories on the frequency definition of probability. However, from the 1950s there has been a growing body of workers whose methods are based on a personal definition of probability, as presented in this book. Their approach is known as Bayesian Statistics, and it is as a foundation for Bayesian Statistics that this book is primarily intended. Many ideas and methods of Bayesian Statistics, and most of the statistical models and standard distributions used by statisticians generally, lie beyond the scope of this book.

An important area of application of personal probabilities is Decision Theory. Major decisions are nearly always taken in a context of uncertainty about their consequences. Measurement of these uncertainties is a key factor in making decisions. The other key factor is to measure the values of those consequences. These measurements are called utilities. Probabilities and utilities are complementary in such a way that it is not possible to give a complete account of one without the other. Utilities were mentioned in Section 1.3, when discussing the role of bets in determining probabilities. The whole matter of utilities and Decision Theory lies outside this book.

The reader will therefore see that he or she has only made a beginning and may, I hope, then be tempted to delve more deeply into these large and intriguing fields of study.

Appendix: Solutions to exercises

1(a)

1. to 6. Clearly you should use your own probability measurements here. However, you should reconsider Exercises 4 to 6 again after Chapter 8.

1(b)

1. $P(W \wedge A | H) = 1/(1+20)$ and $P(W \wedge R | H) = 1/(1+15)$. Now notice that $W = (W \wedge A) \vee (W \wedge R)$, and that $(W \wedge A) \wedge (W \wedge R)$ is logically impossible, so we can use the Addition Law with $E = W \wedge A$ and $F = W \wedge R$. So $P(W | H) = 1/21 + 1/16 = 37/336$.

2. and 3. Use your own probabilities again. For Exercise 2 use $P(E_{10} | H) = P(F_{11} | H) + P(F_{12} | H)$. Then for Exercise 3, $P(E_9 | H) = P(F_{10} | H) + P(E_{10} | H)$, et cetera.

4. Let $P(E | H) = f \{O(E | H)\}$. Then the Addition Law holds if and only if for all non-negative x and y

$$f\left(\frac{xy-1}{x+y+2}\right) = f(x) + f(y). \tag{A.1}$$

Our other constraints are $f(0) = 1$ and $f(x)$ is decreasing as x increases. Setting $x = y$ in (A.1) gives $f((x-1)/2) = 2f(x)$. Applying this repeatedly, starting at $x = 1$, gives

$$f(1) = 1/2, \ f(3) = 1/4, \ f(7) = 1/8, \ f(15) = 1/16, \ldots$$

Using these numbers, and combining them using (A.1), we can find the inverse function $f^{-1}(t)$ for all t in $(0, 1]$. E.g. $3/4 = f(1) + f(2) = f(1/3)$. Since $f(x)$ is decreasing, this procedure implicitly determines $f(x)$ uniquely.

1(c)

1. (i) $P(R_1 \wedge R_2 | H) = P(R_1 | H) P(R_2 | R_1 \wedge H)$. By the definition of probability through direct measurement, $P(R_1 | H) = 3/4$, $P(R_2 | R_1 \wedge H) = 2/3$.
$\therefore P(R_1 \wedge R_2 | H) = 1/2$.
(ii) Similarly, $P(R_1 \wedge W_2 | H) = 3/4 \times 1/3 = 1/4$.
(iii) $P(W_1 \wedge R_2 | H) = 1/4 \times 1 = 1/4$.
(iv) $P(W_1 \wedge W_2 | H) = 1/4 \times 0 = 0$.
(v) Then $P(R_2 | H) = P(R_2 \wedge R_1 | H) + P(R_2 \wedge W_1 | H) = 1/2 + 1/4 = 3/4$.
(vi) $P(W_2 | H) = 1 - P(R_2 | H) = 1/4$.
(vii) By the Multiplication Law, $P(R_1 | R_2 \wedge H) = P(R_1 \wedge R_2 | H)/P(R_2 | H) =$

$(1/2)/(3/4)=2/3.$
(viii) Similarly, $P(W_1|R_2 \wedge H)=(1/4)/(3/4)=1/3.$
(ix) $P(R_1|W_2 \wedge H)=1.$

2. Letting $R_1 \equiv$ 'i-th ball is red', $P(E_1|H)=P(R_1|H)=1/2.$
$P(E_2|H)=P(R_1 \wedge R_2|H)=P(R_1|H)P(R_2|R_1 \wedge H)=1/2 \times 2/3=1/3.$ Noting that $E_i=R_i \wedge E_{i-1}$ we find successively that $P(E_3|H)=1/4$, $P(E_4|H)=1/5,\ldots,$ $P(E_{10}|H)=1/11.$

3. Your own probabilities. Use $P(L_1 \wedge L_2|H)=P(L_1|H)P(L_2|L_1 \wedge H).$ The second probability should be relatively high, because if snow falls on the 1st of January it is quite likely to snow again on 2nd January.

4. Even a slight knowledge of U.S. politics suggests that D and B are not independent. By U.S. laws, N is impossible. Knowing this means that if E is any proposition at all then $E \wedge N$ is also impossible, therefore $P(E \wedge N|H)=0=P(E|H)P(N|H).$ Therefore, in a trivial sense, an impossible proposition is independent of any other proposition. Knowing the truth of S would not noticeably change my beliefs about D or B. Therefore S and D are independent given my information, as are S and B.

5. Only if one of them is impossible. See the previous solution.

6. Answers only: (i) $1/3$, (ii) $1/6$, (iii) $1/6$, (iv) $1/3$, (v) $1/2$, (vi) $1/2$, (vii) $2/3$, (viii) $1/3$, (ix) $1/3$.

2(a)

1. If you remember the value you gave in Exercise 1(a)3, try to think about it in a different way this time.

2. The check for coherence is that $P(E_1|H)=1$, since E_1 is certainly true. If you have arrived at $P(E_1|H)>1$ then you have generally measured your $P(F_i|H)$ values too high, and conversely if $P(E_1|H)<1.$

3. It is not easy to think of this probability, because you are not likely to be uncertain about the result of the first draw and yet know the result of the second. Direct measurement is therefore rather inaccurate for most people. Using the Multiplication Law we find that $P(R_1 \wedge R_2|H)=P(R_1|H)P(R_2|R_1 \wedge H)=5/8 \times 6/9=10/24.$ Similarly, $P(R_2|H)=P(R_1 \wedge R_2|H)+P((\neg R_1) \wedge R_2|H)=5/8.$ Then $P(R_1|R_2 \wedge H)=2/3$, and this is clearly a highly accurate measurement.

4. The first and third measurements are non-coherent: the only measurement of $P(B|H)$ which coheres with the first is 0.3, but from the third I should have $P(B|H) \geq 0.35.$

2(b)

1. $P(E_1|H) = P(F_2|H) + P(F_3|H) + ... + P(F_{12}|H)$.

2.
$$P(R_1|R_2 \wedge H) = \frac{P(R_1|H)P(R_2|R_1 \wedge H)}{P(R_1|H)P(R_2|R_1 \wedge H) + P(W_1|H)P(R_2|W_1 \wedge H)}.$$

3. $P(C_2|H) = P(E_1|H)P(E_2|E_1 \wedge H)P(E_3|E_1 \wedge E_2 \wedge H)$, where $E_i \equiv$ 'i-th toss heads'. Measure all three components as $1/2$. Therefore $P_1(C_2|H) = 1/8$. This is highly accurate.

4. $A = (S_1) \vee ((\neg S_1) \wedge (\neg S_2) \wedge S_3)$, and these two are mutually exclusive.
$\therefore P(A|H) = P(S_1|H) + P((\neg S_1) \wedge (\neg S_2) \wedge S_3|H) =$
$P(S_1|H) + P(S_3|(\neg S_1) \wedge (\neg S_2) \wedge H)\{1 - P(S_2|(\neg S_1) \wedge H)\}\{1 - P(S_1|H)\}$.
Similarly, $P(B|H) = P((\neg S_1) \wedge S_2|H) + P((\neg S_1) \wedge (\neg S_2) \wedge (\neg S_3)|H) =$
$P(S_2|(\neg S_1) \wedge H)\{1 - P(S_1|H)\} + \{1 - P(S_3|(\neg S_1) \wedge (\neg S_2) \wedge H)\} \times$
$\{1 - P(S_2|(\neg S_1) \wedge H)\}\{1 - P(S_1|H)\}$.
Expand both formulae and add to show $P(A|H) + P(B|H) = 1$.

5. Use your own value for p. Measurement of $P(F|H)$ directly is difficult because it is rather a small probability.

2(c)

1. Drawing cards is directly analogous to taking balls from a bag. Therefore the following probabilities are obtained: $P(S|H) = 13/52 = 1/4$, $P(A|H) = 4/52 = 1/13$, $P(S \vee A|H) = 16/52 = 4/13$, $P(S \wedge A|H) = 1/52$. To confirm coherence we check that $P(S \vee A|H) = P(S|H) + P(A|H) - P(S \wedge A|H)$.

2. $P_h(F \vee S|H) = 0.2 + 0.2 - 0.1 = 0.3$.

3. By direct measurement, P (number is $i|H) = i/m$, where m is the total number of balls in the bag. $m = 6n(6n+1)/2 = 3n(6n+1)$. The number of balls having even numbers is $2+4+...+6n = 2(1+2+...+3n) = 2 \times 3n(3n+1)/2 = 3n(3n+1)$. $\therefore P(A|H) = \{3n(3n+1)\}/\{3n(6n+1)\} = (3n+1)/(6n+1)$. Similarly, we find that $P(C|H) = (2n+1)/(6n+1)$ and $P(A \wedge C|H) = (n+1)/(6n+1)$.
$\therefore P(A \vee C|H) = \{(3n+1)+(2n+1)-(n+1)\}/(6n+1) = (4n+1)/(6n+1)$.
Next, $P((\neg A) \vee (\neg C)|H) = 1 - P(A \wedge C|H) = 5n/(6n+1)$.
Finally, $P(A \vee (\neg C)|H) = P(A|H) - P(A \wedge C|H) = 2n/(6n+1)$.

2(d)

1. $P(L|H) = 0.10 + 0.14 + 0.16 = 0.4$, $P(U|H) = 0.16 + 0.15 + 0.20 + 0.22 = 0.76$.
$P(E|H) = 0.14 + 0.18 + 0.22 = 0.54$. $\therefore P(O|H) = 1 - 0.54 = 0.46$.

2. $P(E_1|H) = 0.11 + 0.10 + 0.08 + 0.04 = 0.33$.
$P(E_2|H) = 0.02 + 0.01 + 0.01 + 0.01 = 0.05$.

$P(E_3|H) = 0.12 + 0.10 + 0.02 + 0.01 = 0.25.$
$P(E_4|H) = 0.11 + 0.08 + 0.04 + 0.06 + 0.02 + 0.01 = 0.32.$
$P(E_5|H) = 0.10 + 0.05 + 0.04 = 0.19.$

3. There are many ways to check for coherence here. Denote your information by S, since H is one of the propositions defined in the question. An obvious check is $P(D|S) + P(K|S) = 1$. Also $P(K|S) = P(G|S) + P(H|S)$. Another is $P(A|S) + P(J|S) + P(M|S) + P(C|S) = 1$. The conjunction inequality is another source of coherence checks. E.g. did you have $P(A|S) \geq P(I|S)$?

4. $P(E \vee F|H) = P(E|H) + \{P(F|H) - P(E \wedge F|H)\} \geq P(E|H)$, using the conjunction inequality.

5. The notion of a 'randomly chosen' student suggests measurements such as $P_r(B|H) = 85/170$. The numbers given in the question allow us to deduce that 90 students have read 'The Lord of the Rings', and so $P_r(L|H) = 90/170 = 9/17$.

2(e)

1. In Exercise 2(d)1, $S_1, S_2, \ldots, S_6$ is a logical partition, and $0.10 + 0.14 + 0.16 + 0.18 + 0.20 + 0.22 = 1$. In Exercise 2(d)2, the 16 propositions whose probabilities are given in the table form a partition given the information that the probability of either team scoring more than 3 goals is zero. The 16 probabilities sum to one as required.

2. Let $W \equiv$ 'reads newspaper', $X \equiv$ 'Spain', $Y \equiv$ 'Mexico', $Z \equiv$ 'Spanish-speaking'. Not all combinations of answers are logically possible because X or Y implies Z. Let $E_1 = W \wedge X \wedge Y \wedge Z$, $E_2 = W \wedge X \wedge (\neg Y) \wedge Z$, $E_3 = W \wedge (\neg X) \wedge Y \wedge Z$, $E_4 = W \wedge (\neg X) \wedge (\neg Y) \wedge Z$, $E_5 = W \wedge (\neg X) \wedge (\neg Y) \wedge (\neg Z)$, $E_6 = (\neg W) \wedge X \wedge Y \wedge Z$, $E_7 = (\neg W) \wedge X \wedge (\neg Y) \wedge Z$, $E_8 = (\neg W) \wedge (\neg X) \wedge Y \wedge Z$, $E_9 = (\neg W) \wedge (\neg X) \wedge (\neg Y) \wedge Z$, $E_{10} = (\neg W) \wedge (\neg X) \wedge (\neg Y) \wedge (\neg Z)$.
Then $A = W = E_1 \vee E_2 \vee E_3 \vee E_4 \vee E_5$, $B = X \wedge (\neg Y) = E_2 \vee E_7$, et cetera. This may help you to formulate coherence checks like those given in the solution 2(d)3 above. However, the whole system comprises only 13 equations in 10 unknowns, so there are only 3 linearly independent checks. You may find others but they will all be implied by combinations of the three given in solution 2(d)3.

3. P (January) $= 31/365$, since there are 31 days in January. P (February)$= 28/365$, ignoring leap years, and so on. The final part depends on your birthday.

4. Define the partition $G_1 \equiv F_1$, $G_1 \equiv E_2$, $G_3 \equiv F_2 \wedge \neg E_2$, $G_4 \equiv E_3 = F_3$. The remaining propositions can be expressed as $E_1 = G_1 \vee G_2$ and $F_2 = G_2 \vee G_3$. It is easy to check that the two sets cohere, i.e. all 6 probabilities hold, if we set $P(G_1|H) = P(G_2|H) = P(G_4|H) = 1/3$ and $P(G_3|H) = 0$. This seems unreasonable. Although it is possible to hold both beliefs, it is more natural to hold neither, since the natural equi-probable partition is the G_i s.

3(a)

1. $P(C|H) = P(D_{11}|H)P(C|D_{11}\wedge H) + P(D_{12}|H)P(C|D_{12}\wedge H) + P(D_{13}|H) \times P(C|D_{13}\wedge H)$. We have objective probabilities $P(C|D_{11}\wedge H) = {}^{11}/50$ and so on. Therefore, applying the above elaboration we have $P_1(C|H) = 0.1 \times ({}^{11}/50) + 0.3 \times ({}^{12}/50) + 0.6 \times ({}^{13}/50) = {}^1/4$.

2. Extending the argument, $P(D \wedge S|H) = P(G|H)P(D|G \wedge H) \times P(S|G \wedge H) + P(\neg G|H)P(D|(\neg G)\wedge H)P(S|(\neg G)\wedge H)$. The elaborated measurement is $P_a(D \wedge S|H) = 0.8 \times 0.6 \times 0.5 + 0.2 \times 0.2 \times 0.3 = 0.252$. By similar elaborations, $P_a(D|H) = 0.52$, $P_a(S|H) = 0.46$.

3. Use $P(V_1|H) = {}^3/10$, $P(V_2|H) = {}^1/10$, et cetera. Extending the argument, $P(D|H) = 0.4$.

3(b)

1. (i) $P(L_B|L_A \wedge H) = P(L_A \wedge L_B|H)/P(L_A|H)$. Therefore the measured probability is $P_1(L_B|L_A \wedge H) = {}^2/3$.
(ii) Using also $P(L_B \wedge (\neg L_A)|H) = P(L_B|H) - P(L_A \wedge L_B|H)$ (Addition Law), we find $P_1(L_B|(\neg L_A)\wedge H) = \{({}^5/8) - ({}^1/4)\}/\{1 - ({}^3/8)\} = 0.6$.

2. Let $G \equiv$ 'healthy', $L \equiv$ 'lymphatic Krupps' and $N \equiv$ 'non-lymphatic Krupps' and $R \equiv$ 'positive result'. The question gives $P(G|H) = 0.9$, $P(L|H) = 0.06$, $P(N|H) = 0.04$, $P(R|L \wedge H) = 1$, $P(R|N \wedge H) = 0.75$, $P(R|G \wedge H) = 0.2$. We require $P(L \vee N|R \wedge H) = 1 - P(G|R \wedge H)$. Using Bayes' theorem, $P(G|R \wedge H) = (0.9 \times 0.2)/(0.9 \times 0.2 + 0.06 \times 1 + 0.04 \times 0.75) = {}^2/3$. $\therefore P(L \vee N|R \wedge H) = {}^1/3$. (You could alternatively use Bayes' theorem to obtain $P(L|R \wedge H) = {}^2/9$ and $P(N|R \wedge H) = {}^1/9$ and then apply the Addition Law.)

3. Let $A \equiv$ 'Atkins paroled' and define B, C similarly. Let $I \equiv$ 'Governor names Brown'. Then $P(A|H) = P(B|H) = P(C|H) = {}^1/3$, $P(I|A \wedge H) = {}^1/2$, $P(I|B \wedge H) = 0$, $P(I|C \wedge H) = 1$. By Bayes' theorem, $P(A|I \wedge H) = {}^1/3$, so Governor is correct. Similarly, $P(C|I \wedge H) = {}^2/3$, so Carp is wrong also.

4(a)

1. Using the product theorem, the required probability is $({}^9/12) \times ({}^8/11) \times ({}^7/10) \times ({}^6/9) = {}^{14}/55$.

2. Product theorem again. Probability is $({}^1/2) \times ({}^2/3) \times ({}^3/4) \times \cdots$. The last term is $n/(n+1)$. Everything cancels to leave $1/(n+1)$.

3. Your personal probabilities are required. Use the product theorem as an elaboration and measure the component probabilities. Note, however, that exchange rates tend to move in the same direction on successive days. Therefore your probability for a fall on the second day, knowing that there was a fall on the first day, should be higher than your probability for a fall on the first day.

4. Using mutual independence, $P(F|H) = \prod_{i=1}^{10} P(E_i|H)$, and each term in the product is p.

5. Let $T \equiv$ 'bell triggered', then $\neg T = \{(F_1 \wedge F_2) \vee (F_3 \wedge F_4)\} \vee \{F_5 \vee (F_6 \wedge F_7 \wedge F_8)\}$. Using the disjunction theorem for the probabilities of the propositions in curly braces, and the Multiplication Law (remembering that all F_is are independent), we find $P(T|H) = 1 - (2a^2 - a^4)(a + a^3 - a^4)$. For $a = 0.5, 0.1, 0.01$ we have $P(T|H) = 0.754, 0.998, 1.000$.

6. For instance, independence implies $P(L|S) = P(A|S)P(G|S)$. In fact, independence as suggested means that $P(E_1|S), \ldots, P(E_{10}|S)$ (solution 2(e)2) can be expressed in terms of six probabilities, so you should be able to find four distinct independence tests.

4(b)

1. For example $\binom{5}{2} = 5!/(2!3!)$, which equals $5 \times 4 \times 3/(3 \times 2) = 10$. The values for $k = 0, 1, 3, 4, 5$ are similarly found to be 1, 5, 10, 5, 1.

2. $10!/(5!2!3!) = 10 \times 9 \times 8 \times 7 \times 6/(2 \times 3 \times 2) = 2520$.

3. $\binom{8}{5} = 56$.

4.
$$\frac{n!}{(k-1)!\,(n-k+1)!} + \frac{n!}{k!\,(n-k)!} = \frac{n!}{k!\,(n-k+1)!}(k+(n-k+1))$$
$$= \frac{(n+1)!}{k!\,(n-k+1)!}.$$

5. $\left(_{4,\,3,\,3,\,2}^{12}\right) = 277200$.

6. Treat the four empty squares as another make of car. Then the answer is $\left(_{4,\,3,\,3,\,2,\,4}^{16}\right) = 504504000$.

4(c)

1. Let $S_i \equiv$ 'success with i-th die'. The S_is are mutually independent with $P(S_i|H) = 0.5$. Therefore they comprise 12 trials. Let $R_j \equiv$ 'j successes', and the R_is are logically mutually exclusive. $P(R_6|H) = \binom{12}{6} 0.5^6 0.5^6 = 0.226$. Let $L \equiv$ 'less than three successes' $= R_0 \vee R_1 \vee R_2$. Therefore $P(L|H) = P(R_0|H) + P(R_1|H) + P(R_2|H)$. We find $P(L|H) = 0.0002 + 0.0029 + 0.0190 = 0.0221$.

2. We have 15 trials with probability 0.6 of success at each trial. Therefore the probability of 10 successes is $\binom{15}{10} 0.6^{10} 0.4^5 = 0.186$.

3. We have 8 trials, probability 0.75 of positive at each trial. Probability of 6 or more is $\binom{8}{6} 0.75^6 0.25^2 = 0.311$.

4. Given n binary trials with probability p of success at each, let $R_j \equiv$ 'j successes'. Then the R_js form a partition, so that $1 = \sum_{j=1}^{n} P(R_j | H)$. The binomial probabilities theorem completes the proof.

5. $\binom{10}{4} 0.1^4 0.9^6 = 0.011$.

4(d)

1. Four four-way trials with probabilities 0.42, 0.28, 0.18, 0.12 on each partition. $P(S(1,1,1,1)|H) = \binom{4}{1,1,1,1} 0.42 \times 0.28 \times 0.18 \times 0.12 = 0.061$.

2. 'Total 20'$=S(2,0,6) \vee S(1,2,5) \vee S(0,4,4)$. Add the three resulting probabilities to give $0.012 + 0.097 + 0.054 = 0.163$.

3. The answer is 2 red, 2 white, 1 green. You could guess this, since it matches the proportions in the bag, but to prove it you either require some awkward mathematics or patience in trying all the possible permutations! The probability is 0.1536.

4. Let $E \equiv$ 'same number of 7s as 2s', $F \equiv$ 'more 2s than 7s'. Then it is natural to judge $P(D|H) = P(F|H)$, and this is easy to justify by symmetry arguments. Since D, E, F form a partition, $1 = P(D|H) + P(E|H) + P(F|H)$ $= P(E|H) + 2P(D|H)$. Therefore $P(D|H) = (1 - P(E|H))/2$. Now $E = S(0,0,4) \vee S(1,1,2) \vee S(2,2,0)$. Calculating these three probabilities, and summing to give $P(E|H)$, we obtain $P(D|H) = 0.256$.

4(e)

1. Let $A \equiv$ 'first is red', $B \equiv$ 'third is white' and $C \equiv$ 'sixth is green'. Then $P(A \wedge B \wedge C | H) = P(A|H) P(B|A \wedge H) P(C|A \wedge B \wedge H) = (5/11) \times (4/10) \times (2/9) = 4/99$.

2. There are 8 Hearts in the 42 unseen cards, so probability is $8/42 = 4/21$.

3. Assuming that you give equal probabilities to all possible sequences, we have a complete deal of 5 from $(3,2)$. Then the required probability is $(2/5) \times (1/4) = 0.1$.

4(f)

1. We have a deal of 7 from $(11, 24, 15)$. $P(S(2,3,2)|H) =$ $\binom{11}{2} \binom{24}{3} \binom{15}{2} / \binom{50}{7} = 0.117$.

2. In getting to B she must make 5 moves North and 8 East. She passes through C if in her first 5 moves she goes North three times and East twice. So consider these first 5 moves as a deal of 5 from $(5,8)$, and the required probability is $P(S(3,2)|H) = \binom{5}{3} \binom{8}{2} / \binom{13}{5} = 0.218$.

3. A deal of 13 from $(13, 39)$. Probability is $P(S(5, 8)|H) = 0.0054$.

4(g)

1. For general k we have $P(S(2, 3)|H) = \binom{k}{2}\binom{2k}{3}/\binom{3k}{5}$ For $k = 2, 5, 10$ we obtain $2/3, 0.3996, 0.3600$. As $k \to \infty$ we reach the binomial probability $\binom{5}{2}(1/3)^2(2/3)^3 = 0.3292$.

2. The multinomial probability is $\binom{7}{2, 3, 2} 0.22^2 0.48^3 0.3^2 = 0.101$. This is the limit of the probability in 4(f)1 as the number of students increases whilst proportions in the various faculties are fixed.

5(a)

1. T, L, S and P are all defined on the partition E_0, E_1, E_2, E_3, E_4. We find $E_0 = {}'T = 0' = {}'L = 5' = {}'S = 5' = {}'P = 0'$.
$\therefore P(T = 0|H) = P(L = 5|H) = P(S = 5|H) = P(P = 0|H) = 0.25$. The four distributions are given in the following table.

T	0	1	2	3	4
L	5	6	7	8	9
S	5	4	3	2	1
P	0	$2/6$	$4/7$	$6/8$	$8/9$
Prob	0.25	0.32	0.27	0.13	0.03

2. $P(G = 0|H) = 0.12 + 0.15 + 0.10 + 0.05 = 0.42$. Similarly $P(G = 1|H) = 0.33$, $P(G = 2|H) = 0.17$, $P(G = 3|H) = 0.08$.

3. The joint distribution has probabilities $P((X = x) \wedge (Y = y)|H) = P(X = x|H)P(Y = y|H)$ because they are independent. Therefore

y / x	0	1	2	3	4	5	6
0	.0600	.0800	.0800	.0600	.0600	.0400	.0200
1	.0450	.0600	.0600	.0450	.0450	.0300	.0150
2	.0375	.0500	.0500	.0375	.0375	.0250	.0125
3	.0075	.0100	.0100	.0075	.0075	.0050	.0025

$P(X + Y = 2|H) =$
$P((X = 0) \wedge (Y = 2)|H) + P((X = 1) \wedge (Y = 1)|H) + P((X = 2) \wedge (Y = 0)|H) =$
$0.08 + 0.06 + 0.0375 = 0.1775$. Similarly we find

z	0	1	2	3	4
$P(Z = z|H)$	0.06	0.125	0.1775	0.1775	0.165

z	5	6	7	8	9
$P(Z = z|H)$	0.1325	0.095	0.0475	0.0175	0.0025

5(b)

1. $P(X=0|H)=P(X=0|H\wedge E)P(E|H)+P(X=0|H\wedge(\neg E))P(\neg E|H)$.
$\therefore P_3(X=0|H)=0.6\times0.7+0.3\times0.3=0.51$. Similarly $P_3(X=1|H)=0.3$ (note the irrelevant information theorem), $P_3(X=2|H)=0.13$, $P_3(X=3|H)=P_3(X=4|H)=0.03$.

2. $P((X=1)\wedge(Y=1)|H)=P(X=1|H)P(Y=1|(X=1)\wedge H)=0.06\times0.5$ since $Y=1$ is a correct count when $X=1$. The table shows *part* of the joint distribution.

x	1	2	3	4
y				
0	0.015	0.0	0.0	0.0
1	0.03	0.0425	0.0	0.0
2	0.015	0.085	0.0475	0.0
3	0.0	0.0425	0.095	0.045
4	0.0	0.0	0.0475	0.09
5	0.0	0.0	0.0	0.045

Summing across rows of the table gives the marginal distribution of Y.

y	0	1	2	3	4
$P_2(Y=y|H)$	0.015	0.0725	0.1475	0.1825	0.1775

z	5	6	7	8	9
$P_2(Y=y|H)$	0.16	0.13	0.08	0.03	0.005

$P(X=1|H\wedge(Y=2))=P((X=1)\wedge(Y=2)|H)/P(Y=2|H)=0.015/0.1475=0.102$.
Similarly $P(X=2|H\wedge(Y=2))=0.576$, $P(X=3|H\wedge(Y=2))=0.322$ and the remaining values of X are impossible given $H\wedge(Y=2)$.

3. Given $H_1\wedge(N=20)$, Y is a deal of 2 from $(15,5)$. Thus
$P(Y=y|H_1\wedge(N=20))=\binom{15}{y}\binom{5}{2-y}/\binom{20}{2}$. Then $P(Y=0|H_1\wedge(N=20))=$
$1\times10/190=0.05263$, $P(Y=1|H_1\wedge(N=20))=15\times5/190=0.39474$,
$P(Y=2|H_1\wedge(N=20))=105\times1/190=0.55263$. $P(Y=y|H)$ is obtained by extending the argument, using the second column of Table 5.11 and the conditional distributions of Y given $H_1\wedge(N=n)$ for $n=15,16,\ldots,30$. These are computed as for the case $N=20$ already found. Eventually we obtain

y	0	1	2
$P(Y=y|H_1)$	0.0867	0.4146	0.4987

4. By Bayes' theorem we find
$P(N=n|H_1\wedge(Y=0))=P(N=n|H_1)P(Y=0|H_1\wedge(N=n))/P(Y=0|H_1)$.
For instance, $P(N=20|H_1\wedge(Y=0))=0.083\times0.05263/0.0867=0.079$.
The full distribution is

n	15	16	17	18	19	20
$P(N=n\mid H_1\wedge(Y=0))$	0	0	0.003	0.016	0.045	0.079

n	21	22	23	24	25	26
$P(N=n\mid H_1\wedge(Y=0))$	0.107	0.126	0.140	0.135	0.104	0.078

n	27	28	29	30
$P(N=n\mid H_1\wedge(Y=0))$	.065	.048	.026	.028

5(c)

1. $E(X\mid H)=0\times0.4+1\times0.3+2\times0.25+3\times0.05=0.95$. $E(Y\mid H)=0\times0.15+1\times0.2+...+6\times0.05=2.45$. $E(Z\mid H)=0\times0.06+1\times0.125+...+9\times0.0025=3.4$ $(=0.95+2.45)$.

2. $var(X\mid H)=(0-0.95)^2\times0.4+(1-0.95)^2\times0.3+(2-0.95)^2\times0.25+(3-0.95)^2\times0.05=0.8475$. $\therefore sd(X\mid H)=0.9206$.
$var(Y\mid H)=(0-2.45)^2\times0.15+...+(6-2.45)^2\times0.05=3.0475$, $sd(Y\mid H)=1.7457$.
$var(Z\mid H)=(0-3.4)^2\times0.06+...+(9-3.4)^2\times0.0025=3.895$, $sd(Z\mid H)=1.9736$.

3. $E(X\mid H)=1\times0.06+2\times0.17+...+8\times0.02=4.05$.
$E(X\mid H\wedge(Y=2))=1\times0.102+2\times0.576+3\times0.322=2.22$.
$var(X\mid H)=(1-4.05)^2\times0.06+...+(8-4.05)^2\times0.02=3.1675$.
$\therefore sd(X\mid H)=1.78$. $var(X\mid H\wedge(Y=2))=0.3756$. $\therefore sd(X\mid H\wedge(Y=2))=0.613$ (<1.78).

4. $E(T\mid H)=1.37$, $E(L\mid H)=6.37$, $E(S\mid H)=3.63$, $E(P\mid H)=0.3852$.
$sd(T\mid H)=sd(L\mid H)=sd(S\mid H)=1.083$, $sd(P\mid H)=0.2686$.

5(d)

1. The distribution is unimodal with mode at $Y=3$. Its mean is found to be $E_2(Y\mid H)=4.05$ and its standard deviation $sd_2(Y\mid H)=1.915$. It is slightly skewed to the right.

2. (5.18) implies $P(X=x\mid H)=P(X=x\mid H\wedge E)$ for all x. Extending the argument, $P(X=x\mid H)=P(X=x\mid H\wedge E)P(E\mid H)+P(X=x\mid H\wedge(\neg E))P(\neg E\mid H)$
$\therefore P(X=x\mid H)\{1-P(E\mid H)\}=P(X=x\mid H\wedge(\neg E))P(\neg E\mid H)$
$\therefore P(X=x\mid H\wedge(\neg E))=P(X=x\mid H)$ for all x. By Bayes' theorem
$P(E\mid H\wedge(X=x))=P(E\mid H)P(X=x\mid H\wedge E)/P(X=x\mid H)=P(E\mid H)$.

3. $\Sigma_x P(\text{Po}(m)=x)=\Sigma_{x=0}^{\infty}m^x\,e^{-m}/x!=\left[\Sigma_{x=0}^{\infty}m^x/x!\right]e^{-m}=e^m\,e^{-m}=1$.
$E(\text{Po}(m))=\Sigma_{x=0}^{\infty}x\,m^x\,e^{-m}/x!=0\times m+m\,\Sigma_{x=1}^{\infty}m^{x-1}\,e^{-m}/(x-1)!=m\times1=m$.

4. Your own probabilities are required. Successive months are not independent. Therefore, when comparing with the binomial distribution you should find that you have given higher probabilities to $S=0$ and $S=12$ than the corresponding binomial probabilities.

5. $P(Y=0|H)=P(X=1|H)=P(right\ first\ time|H)=1/6$. $P(Y=1|H)=$
$P(wrong\ first\ time\ but\ right\ second\ time|H)=(5/6)(1/6)=5/36$. In general,
$P(Y=y|H)=(5/6)^{y-1}(1/6)$. $\therefore Y|H \sim Ge(1/6)$. $\therefore E(Y|H)=(1-1/6)/(1/6)=5$.
Since X is one more than Y we have $E(X|H)=6$ (see Section 6.3).

In the second formulation $P(X=x|H)=P(first\ lever\ tried\ is\ x\ places\ from$
$correct\ one|H)=1/6$. $\therefore X|H \sim U(6)$. $\therefore E(X|H)=3.5$. The mouse expects to
take fewer attempts under this method.

6. Your own distributions are required. You may have considerable knowledge
of some of these. For instance, a chemist may give C the degenerate distribu-
tion $P_1(C=55|H)=1$. A scholar of music might have a very high probability
for $S=9$. Lesser mortals will find the Poisson and negative binomial families
most useful here.

7. By extending the argument $P(W|H)=\sum_{n=0}^{\infty}P(W|H\wedge(N=n))P(N=n|H)$.
You have measured probabilities $P_1(N=n|H)$ from the Poisson distribution and
objective probabilities $P(W|H\wedge(N=n))=1/(n+1)$.

$$\therefore P_1(W|H) = \sum_{n=0}^{\infty} 100^n\ e^{-100}/(n+1)! = 0.01 \times \sum_{n=0}^{\infty} 100^{n+1} e^{-100}/(n+1)!$$

$$= 0.01 \times (1-e^{-100}) = 0.01\ .$$

6(a)

1.

$$P((R=r)\wedge(X=x)|H) = \binom{N}{r}p^r(1-p)^{N-r}\binom{r}{x}\binom{N-r}{n-x}/\binom{N}{n}$$

$$= p^r(1-p)^{N-r}\frac{n!(N-n)!}{x!(r-x)!(n-x)!(N-n+x-r)!}\ .$$

To obtain $P(X=x|H)$ we sum this joint distribution over the possible values of
R. Now the hypergeometric distribution of X means that $x \leq r$ and $x \geq n+r-N$,
therefore the range of r is $x \leq r \leq x-n+N$.

$$P(X=x|H) = \sum_{r=x}^{x-n+N} P((R=r)\wedge(X=x)|H)$$

$$= \sum_{r^*=0}^{N^*} p^{r^*+x}(1-p)^{N^*-r^*+n-x}\binom{n}{x}\binom{N^*}{r^*}$$

$$= \binom{n}{x}p^x(1-p)^{n-x}\ ,$$

where $r^*=r-x$ and $N^*=N-n$. Dividing the joint distribution by this marginal
distribution gives

$$P(R=r|H\wedge(X=x)) = \binom{N-n}{r-x}p^{r-x}(1-p)^{N-n}\ ,$$

so that $R-x|H \wedge (X=x) \sim \text{Bi}(N-n,p)$.

2. Your distribution for R is binomial so think of R as the number of successes in N trials. Your conditional distribution of X given R means that we can think of X as being the number of successes in a deal of n from $(R,N-R)$, so these are the number of successes in n of the original N trials. Therefore X is also binomial. When $X=x$ is known then $R-x$ is the number of successes in the other $N-n$ trials.

3. If $M|H \sim \text{Po}(4)$ then $E(M|H)=4$. Your expected number of arrivals in thirty seconds is half of this, i.e. 2. Since arrivals are a Poisson process, $T|H \sim \text{Po}(2)$.

6(b)

1. It is true that $T=5-S$, $L=S+2T$ and $P=2T/L$. From (6.7) we know that $E(T|H)=5-E(S|H)$ and $E(L|H)=E(S|H)+2E(T|H)$. From (6.7) and (6.10), $E(P|H)E(L|H)=2E(T|H)$ if P and L are independent given H. However, it is clear that in fact they are not independent.

2. Let X_1, X_2, X_3, X_4 be the scores on the four individual dice, so that $F=\Sigma_{i=1}^{4}X_i$. Now $X_i|H \sim \text{U}(6) \therefore E(X_i|H)=3.5$ for $i=1,2,3,4$. $\therefore E(F|H)=\Sigma_{i=1}^{4}E(X_i|H)=14$. We also judge the X_is mutually independent given H, $\therefore var(F|H)=\Sigma_{i=1}^{4}var(X_i|H)=4\times35/12=35/3$.

3. $E(S|H)=E(M|H)/2=2=E(T|H)$. Using (6.15), $var(S|H)=var(M|H)/4=1$, but $var(T|H)=2$. In a Poisson process, the number of arrivals in the first thirty seconds may be different from the number in the following thirty seconds. Therefore the number in thirty seconds (T) is more variable than half the number in one minute (S).

4. $E(X_i|H)=0\times P(X_i=0|H)+1\times P(X_i=1|H)=0\times(1-p_i)+1\times p_i=p_i$. $\therefore E(Y|H)=E(\Sigma_{i=1}^{n}X_i|H)=\Sigma_{i=1}^{n}E(X_i|H)=\Sigma_{i=1}^{n}p_i$. We find $E(X_i^2|H)=0^2\times(1-p_i)+1^2\times p_i=p_i$. Therefore from (6.12), $var(X_i|H)=p_i-p_i^2=p_i(1-p_i)$. If the E_is are independent then the X_is are independent and (6.16) can be used repeatedly to give $var(\Sigma_{i=1}^{n}X_i|H)=\Sigma_{i=1}^{n}var(X_i|H)=\Sigma_{i=1}^{n}p_i(1-p_i)$. If $Y|H \sim \text{Bi}(n,p)$ then Y is the number of successes in n binary trials. We can attach an indicator variable to each trial, with $P(X_i=1|H)=p$ for $i=1,2,\ldots,n$. Trials are independent. We therefore find $E(Y|H)=\Sigma_{i=1}^{n}p=np$, and $var(Y|H)=np(1-p)$.

5. We define an indicator random variable for each partition in a deal of n from $(R,N-R)$. $P(X_i=1|H)=R/N$. $Y=\Sigma_{i=1}^{n}X_i$ has the $\text{Hy}(N,R,n)$ distribution and the result of Exercise 4 allows us to prove $E(\text{Hy}(N,R,n))=nR/N$. (We cannot get the variance so simply because the partitions in a deal are not independent.)

6. $d(a) = E(X^2 - 2aX + a^2 | H) = E(X^2 | H) - 2aE(X | H) + a^2$ using (6.7) and (6.8). Therefore $d(a) = \{a - E(X | H)\}^2 + E(X^2 | H) - \{E(X | H)\}^2$. The first term is the only one involving a and is minimized by letting $a = E(X | H)$.

7. Let $X | H \sim \text{Po}(m)$. $E(X(X-1) | H) = \sum_{x=0}^{\infty} x(x-1) P(X=x | H)$. Notice that $x(x-1)$ is zero at $x=0$ or $x=1$, so we can collapse the sum to

$$\sum_{x=2}^{\infty} x(x-1) P(X=x | H) = \sum_{x=2}^{\infty} x(x-1) m^x e^{-m} / x!$$

$$= \sum_{x=2}^{\infty} m^x e^{-m} / (x-2)!$$

$$= m^2 \sum_{x=2}^{\infty} m^{x-2} e^{-m} / (x-2)! .$$

Defining $x^* = x - 2$ we see that the sum is the sum of probabilities in the Po(m) distribution and therefore equals one. Thus $E(X(X-1) | H) = m^2$. From (6.13), $\text{var}(X | H) = m^2 + m - m^2 = m$.

6(c)

1. Use (6.18). First find $E(N M | H) = 0 \times 0 \times 0.112 +) \times 1 \times 0.089 +$ $...+1 \times 1 \times 0.084 + ... + 4 \times 4 \times 0.005 = 2.219$. We need $E(N | H)$ and $E(M | H)$: $P(N=0 | H) = 0.112 + 0.089 + 0.056 + 0.022 + 0.006 = 0.285$, $P(N=1 | H) = 0.294$, $P(N=2 | H) = 0.234$, $P(N=3 | H) = 0.131$, $P(N=4 | H) = 0.056$. Notice the symmetry in the table – M has the same marginal distribution. We find $E(N | H) = E(M | H) = 1.379$. $\therefore \text{cov}(N, M | H) = 2.219 - 1.379^2 = 0.31736$. We also find $\text{var}(N | H) = \text{var}(M | H) = 1.40336$. $\therefore \text{corr}(N, M | H) = 0.31736 / 1.40336 = 0.226$.

2. In Exercise 5(c)2 we found $E(X | H) = 4.05$, $sd(X | H) = 1.78$. In 5(b)2 we found the marginal distribution of Y, from which we obtain $E(Y | H) = 4.05$, $sd(Y | H) = 1.915$. From the joint distribution in 5(b)2, $E(XY | H) = 1 \times 0 \times 0.015 + 1 \times 1 \times 0.03 + ... + 8 \times 9 \times 0.005 = 19.57$. $\therefore \text{cov}(X, Y | H) = 19.57 - 4.05^2 = 3.1675$. $\therefore \text{corr}(X, Y | H) = 3.1675 / (1.78 \times 1.915) = 0.929$.

3. From the definition (6.23), if $\text{corr}(X, Y | H) > 0$ then $\text{cov}(X, Y) > 0$ because standard deviations are always positive. Now use (6.17). $\text{var}(Z | H)$ will increase if $\text{cov}(X, Y)$ becomes positive instead of zero.

4. X and Y are clearly not independent. When X is known then we know Y. Mathematically, $P((X=0) \wedge (Y=0) | H) = P(X=0 | H) = 0.4$, which does not equal $P(X=0 | H) P(Y=0 | H)$. Possible values of $XY = X^3$ are $-8, -1, 0, 1, 8$ with probabilities $0.1, 0.2, 0.4, 0.2, 0.1$ respectively. We find $E(XY | H) = 0$. Since we also have $E(X | H) = 0$, from (6.18) $\text{cov}(X, Y | H) = 0 \therefore \text{corr}(X, Y | H) = 0$. [More generally, if X is symmetric about zero then $XY = X^3$ will also be symmetric about zero, and we will always find $E(XY | H) = E(X | H) = 0$.]

6(d)

1. $E(X|H \wedge (Y=y)) = y + E\{Po(m(1-p))\} = y + m(1-p)$.
$E(Y|H \wedge (X=x)) = E\{Bi(x,p)\} = xp$.
Both regressions are linear. In particular, the coefficient b in (6.36) is 1, which from (6.38) equals $cov(X,Y|H)/var(Y|H)$. Therefore
$cov(X,Y|H) = var(Y|H) = var\{Po(mp)\} = mp$.
Now $var\{E(X|Y \wedge H)|H\} = var(Y + m(1-p)|H) = var(Y|H) = mp$, from (6.15). And $var(X|H) = var(Po(m)) = m$. $\therefore I(X,Y|H) = mp/m = p$. Also $var\{E(Y|X \wedge H)|H\} = var(Xp|H) = p^2 var(X|H) = mp^2$.
$\therefore I(Y,X|H) = mp^2/mp = p$. Finally,
$\{corr(X,Y|H)\}^2 = \{cov(X,Y|H)\}^2/\{var(X|H) var(Y|H)\} = (mp)^2/(m \times mp)$.

2. Direct calculation yields the following table.

m	0	1	2	3	4
$E(N|(M=m) \wedge H)$	1.0211	1.3435	1.5427	1.7252	1.8929

(This is not linear because the increments from each column to the next are unequal.) Now $E\left[\{E(N|(M \wedge H)\}^2|H\right] = 1.0211^2 \times 0.285 + 1.3435^2 \times 0.294 + ... + 1.8929^2 \times 0.056 = 1.9753$. Since $E\left[E(N|H \wedge M)|H\right] = E(N|H) = 1.379$, $var\left[E(N|M \wedge H)|H\right] = 1.9753 - 1.379^2 = 0.0525$. This is close, but not equal, to $\{corr(N,M|H)\}^2 = 0.226^2 = 0.511$. [Notice that because of the symmetry of their joint distribution $I(M,N|H) = I(N,M|H)$, so that the two quadratic information measures may be equal even when the regression is non-linear.]

3.
$E\{cov(X,Y|Z \wedge H)|H\} = E\{E(XY|Z \wedge H) - E(X|Z \wedge H)E(Y|Z \wedge H)|H\} = E(XY|H) - E\{E(X|Z \wedge H)E(Y|Z \wedge H)|H\}$,
$cov\{E(X|Z \wedge H), E(Y|Z \wedge H)|H\} = E\{E(X|Z \wedge H)E(Y|Z \wedge H)|H\} - E\{E(Z|Z \wedge H)|H\}E\{E(Y|Z \wedge H)|H\} = E\{E(X|Z \wedge H)E(Y|Z \wedge H)|H\} - E(X|H)E(Y|H)$. Adding these two together yields $E(XY|H) - E(X|H)E(Y|H) = cov(X,Y|H)$.

4.
$E(B|(N=n) \wedge H) = E(\Sigma_{i=1}^n X_i|(N=n) \wedge H) = \Sigma_{i=1}^n E(X_i|(N=n) \wedge H) = \Sigma_{i=1}^n 1 = n$.
$\therefore E(B|H) = E\{E(B|N \wedge H)|H\} = E(N|H) = 13$.
Also $var\{E(B|N \wedge H)|H\} = var(N|H) = 2.8$. Now the X_is are independent, so $var(B|(N=n) \wedge H) = \Sigma_{i=1}^n var(X_i|(N=n) \wedge H) = \Sigma_{i=1}^n 1 = n$.
$\therefore E\{var(B|N \wedge H)|H\} = E(N|H) = 13$. $\therefore var(B|H) = 2.8 + 13 = 15.8$ using (6.31).

5. Let $X^* = aX + b$, $Y^* = cY + d$. Knowing Y^* is equivalent to knowing Y, therefore $E(X^*|(Y^* = y^*) \wedge H) = aE(X|(Y=y) \wedge H) + b$. $\therefore var\{E(X^*|Y^* \wedge H)|H\} = var\{aE(X|Y \wedge H) + b|H\} = a^2 var\{E(X|Y \wedge H)|H\}$ using (6.15). From (6.15) also, $var(X^*|H) = a^2 var(X|H)$. The required result now follows from the

definition of $I(X,Y|H)$.

7(a)

1. The distribution function is a step function with steps at $x=0,1,2,3,4,5$. $D_{\text{Bi}(5,0.3)}(x)=0$ for $x<0$. At $x=0$ it steps up to $P(\text{Bi}(5,0.3)\leq 0)=$ $P(\text{Bi}(5,0.3)=0)=\binom{5}{0}0.3^0 0.7^5=0.168$, and keeps this value until $x=1$, where it steps up again to $P(\text{Bi}(5,0.3)\leq 1)=P(\text{Bi}(5,0.3)=0)+P(\text{Bi}(5,0.3)=1)=$ $0.168+\binom{5}{1}0.3^1 0.7^4=0.168+0.360=0.528$. Continuing in this way we find the remaining steps: $D_{\text{Bi}(5,0.3)}(2)=0.837$, $D_{\text{Bi}(5,0.3)}(3)=0.969$, $D_{\text{Bi}(5,0.3)}(4)=0.998$, $D_{\text{Bi}(5,0.3)}(5)=1$.

2. $P(S\leq 0.3|H)=D_{S|H}(0.3)=0.3^2=0.09$. $P(0.2<S\leq 0.8|H)=$ $P(S\leq 0.8|H)-P(S\leq 0.2|H)=D_{S|H}(0.8)-D_{S|H}(0.2)=0.8^2-0.2^2=0.6$.

3. Distribution functions are necessarily non-decreasing, with values between 0 and 1. $f_i(x)$ all satisfy these conditions for $x<1$ and $x>2$. Between $x=1$ and $x=2$, $f_1(x)$ rises to a maximum of 0.25 at $x=1.5$ and then decreases. It cannot be a distribution function. Between 1 and 2, $f_2(x)$ increases steadily from $f_2(1)=0$ to $f_1(2)=1$, and it can therefore represent a distribution function. Between 1 and 2, $f_3(x)$ also increases, but $f_3(1)=e-3<0$, so that $f_3(x)$ cannot be a distribution function.

7(b)

1. $D_{Z|H}(z)=\int_{-\infty}^{z}d_{Z|H}(t)\,dt$. For $z<0$, the integrand is always zero, so $D_{Z|H}(z)=0$. Otherwise

$$D_{Z|H}(z)=\int_0^z 2(1+t)^{-3}\,dt=\left[-(1+t)^{-2}\right]_0^z=1-(1+z)^{-2}.$$

$\therefore P(Z>1|H)=1-P(Z\leq 1|H)=1-D_{Z|H}(1)=0.25$.

$$E(Z|H)=\int_{-\infty}^{\infty}z\,d_{Z|H}(z)\,dz=\int_0^{\infty}2z\,(1+z)^{-3}dz$$

$$=2\int_0^{\infty}\{(1+z)^{-2}-(1+z)^{-3}\}\,dz$$

$$=\left[2(1+z)^{-1}\right]_0^{\infty}-\left[(1+z)^{-2}\right]_0^{\infty}=2-1=1.$$

2. Only one value of c will make $d_{Y|H}(y)$ satisfy (7.7). Therefore

$$1=\int_{-\infty}^{\infty}d_{Y|H}(y)\,dy=c\int_0^1 y\,(1-y)\,dy=c\int_0^1(y-y^2)\,dy$$

$$=c\left[y^2/2\right]_0^1-c\left[y^3/3\right]_0^1=c/6.$$

Therefore $c = 6$. We find $D_{Y|H}(y) = 6\int_0^y (t - t^2)\,dt = 3y^2 - 2y^3$ for $0 \le y \le 1$. $D_{Y|H}(y) = 0$ for $y < 0$ and $= 1$ for $y > 1$. $\therefore P(0.4 < y \le 0.6|H) = D_{Y|H}(0.6) - D_{Y|H}(0.4) = 0.648 - 0.352 = 0.296$.

3.

$$E(aX + b|H) = \int_{-bing}^{\infty} (ax + b)\,d_{X|H}(x)\,dx$$

$$= \int_{-\infty}^{\infty} ax\, d_{X|H}(x)\,dx + \int_{-\infty}^{\infty} b\, d_{X|H}(x)\,dx$$

$$= a \int_{-\infty}^{\infty} x\, d_{X|H}(x)\,dx + b \int_{-\infty}^{\infty} d_{X|H}(x)\,dx$$

and use (7.9), (7.7).

4. The density function is non-negative; we need to verify (7.7).
$\int_{-\infty}^{\infty} d_{B|H}(d)\,db = (1/2)\int_{-1}^{1} \pi \cos(\pi b)\,db = (1/2)\left[\sin(\pi b)\right]_{-1}^{1} = (1+1)/2 = 1$.
$P(T \le t|H) = P(3B \le t|H) = P(B \le t/3|H)$. $\therefore D_{T|H}(t) = D_{B|H}(t/3)$.
$\therefore d_{T|H}(t) = (1/3)d_{B|H}(t/3) = \pi\cos(\pi b)/6$ for $-3 \le t \le 3$ (and zero otherwise).
$P(S \le s|H) = P(B^2 \le s|H) = P(-s^{1/2} \le B \le s^{1/2}|H)$.
$\therefore D_{S|H}(s) = D_{B|H}(s^{1/2}) - D_{B|H}(-s^{1/2})$.
$\therefore d_{S|H}(s) = (s^{-1/2}/2)d_{B|H}(s^{1/2}) + (s^{-1/2}/2)d_{B|H}(-s^{1/2}) = s^{-1/2}\cos(\pi s^{1/2})/2$ for
$0 \le s \le 1$, and zero otherwise. $P(E \le f|H) = P(e^B \le f|H) = P(B \le \ln f|H)$.
$\therefore D_{E|H}(f) = D_{B|H}(\ln f)$. $\therefore d_{E|H}(f) = f^{-1}d_{B|H}(\ln f) = (2f^{-1})\cos(\pi \ln f)$ for
$e^{-1} \le f \le e$, and zero otherwise.

5. Since $E(Z|H) = 1$, $var(Z|H) = 2\int_0^{\infty}(z-1)^2(1+z)^{-2}\,dz$. This integral does not exist because the area under the curve $y = (x-1)^2(1+x)^{-2}$ is infinite.

7(c)

1.

$$E(Be(p,q)) = \int_0^1 x\,\{B(p,q)\}^{-1}x^{p-1}(1-x)^{q-1}\,dx$$

$$= \{B(p,q)\}^{-1}\int_0^1 x^p(1-x)^{q-1}\,dx$$

$$= \{B(p,q)\}^{-1}B(p+1,q)$$

$$= \{\Gamma(p+q)\Gamma(p+1)\Gamma(q)\}/\{\Gamma(p)\Gamma(q)\Gamma(p+q+1)\}$$

using (7.27). Then use (7.23), i.e. $\Gamma(p+q+1) = (p+q)\Gamma(p+q)$ and $\Gamma(p+1) = p\,\Gamma(p)$.

2. The estimate gives $E_1(M|H) = 500 = b/a$, using (7.21). Let the range 100 to 1500 represent four standard deviations, i.e. two either side of the mean. (This distribution is evidently quite skew, but we will apply the usual rules of thumb nevertheless.) $\therefore var_1(M|H) = 350^2 = 122500 = b/a^2$.
$\therefore a = 500/122500 = 0.004$. $\therefore b = 500 \times 0.004 = 2$. $\therefore M|H_1 \sim Ga(0.004, 2)$.

3. Use (7.32).
$P(T > 4.7|H) = 1 - P(T \le 4.7|H) = 1 - \Phi(0.09^{-\frac{1}{2}}(4.7 - 4.4)) = 1 - \Phi(1) = 0.1587$
from Table 7.1. $P(4.25 < T \le 5|H) = \Phi((5 - 4.4)/0.3) - \Phi((4.25 - 4.4)/0.3) = \Phi(2) - \Phi(-0.5) = \Phi(2) - (1 - \Phi(0.5)) = 0.6687$.

4. Your own probabilities are required. A beta distribution is most appropriate for W. Either a gamma or a normal distribution is best for T, depending on symmetry. A normal distribution should be adequate for M.

5. $D_{Y|H}(y) = P(Y \le y|H) = P(X \le y(1+y)^{-1}|H) = D_{X|H}(y(1+y)^{-1})$.

$$\therefore d_{Y|H}(y) = (1+y)^{-2} d_{X|H}(y(1+y)^{-1})$$
$$= (1+y)^{-2} \{B(p,q)\}^{-1} y^{p-1}(1+y)^{-(p-1)}(1+y)^{-(q-1)}$$
$$= \{B(p,q)\}^{-1} y^{p-1}(1+y)^{-(p+q)}$$

for $y \ge 0$, and otherwise $d_{Y|H}(y) = 0$.

$$E(Y|H) = E(X(1-X)^{-1}|H) = \int_0^1 x(1-x)^{-1}\{B(p,q)\}^{-1}x^{p-1}(1-x)^{q-1}\,dx$$

$$= \{B(p,q)\}^{-1}B(p+1,q-1) = p/(q-1)$$

using (7.26), (7.27) and (7.23). Similarly, $E(Y^2|H) = p(p+1)/\{(q-1)(q-2)\}$. We therefore obtain $var(Y|H) = p(p+q+1)/\{(q-1)^2(q-2)\}$. If $q \le 2$ then $E(Y^2|H)$ does not exist. (If $q \le 1$ then $E(Y|H)$ does not exist either.)

6. Let $X|H \sim N(0, 1)$ and define $Z = X^2$, then use (7.12).
$d_{Z|H}(z) = z^{-\frac{1}{2}}\{d_{X|H}(z^{\frac{1}{2}}) + d_{X|H}(-z^{\frac{1}{2}})\}/2 = z^{-\frac{1}{2}}(2\pi)^{-\frac{1}{2}}\exp(-z/2)$ using (7.28), for $z > 0$, which is the density of $GA(0.5, 0.5)$. (See (7.20) and remember that $\Gamma(0.5) = \pi^{\frac{1}{2}}$.)

7. Define $X|H \sim N(0, 1)$ and so $Y = \Phi(X)$. Then $D_{Y|H}(y) = P(\Phi(X) \le y|H)$. Consider an instance from Table 7.1 and let $y = 0.8413$. Then $\Phi(X) \le 0.8413$ is equivalent to $X \le 1.0$ and $P(X \le 1|H) = 0.8413$ because $X|H \sim N(0, 1)$. Therefore $D_{Y|H}(0.8413) = 0.8413$. This can be repeated for any other values, and we conclude that $D_{Y|H}(y) = y$ for $0 \le y \le 1$. For $y < 0$, $D_{Y|H}(y) = 0$, and for $y > 1$, $D_{Y|H}(y) = 1$, since Y lies in [0, 1]. This is the distribution function of $Uc(0, 1)$. (In general, for any random variable X and any information H, the same argument shows that $Y \equiv D_{X|H}(X) \sim Uc(0, 1)$.)

8. $S > x$ if and only if every X_i is greater than x. Therefore
$P(S > x|H) = P(\wedge_{i=1}^{n}(X_i > x)|H) = \prod_{i=1}^{n} P(X_i > x|H)$ since the X_is are mutually independent. Now $X_i|H \underset{1}{\sim} Ga(0.2, 1)$, therefore

$$P_1(X_i > x|H) = \int_x^\infty (0.2)\, t^0 e^{-0.2t}\, dt$$

$$= \left[-e^{-0.2t}\right]_x^\infty = e^{-0.2x}$$

$\therefore P(S > x|H) = \exp(-0.2\,nx)$. $\therefore D_{S|H}(x) = 1 - \exp(-0.2\,nx)$.
$\therefore d_{S|H}(x) = 0.2\,n \exp(-0.2\,nx)$ which is the density function of $Ga(0.2n, 1)$.

7(d)

1.

$$\frac{\partial}{\partial k} D_{K,M|H}(k,m) = \begin{cases} km + m^2/2, & \text{if } 0 \le k \le 1 \text{ and } 0 \le m \le 1\,; \\ k + 1/2, & \text{if } 0 \le k \le 1 \text{ and } m > 1\,; \\ 0, & \text{otherwise}\,. \end{cases}$$

$$\therefore d_{K,M|H}(k,m) = \frac{\partial}{\partial m}\frac{\partial}{\partial k} D_{K,M|H}(k,m) = k + m\,,$$

if $0 \le k \le 1$ and $0 \le m \le 1$, and is otherwise zero. $D_{K|H}(k)$ is the limit of $D_{K,M|H}(k,m)$ as $m \to \infty$, i.e. 0 if $k < 0$, $k(k+1)/2$ if $0 \le k \le 1$ and 1 if $k > 1$. Differentiating, $d_{K|H}(k) = k + 1/2$ if $0 \le k \le 1$ and otherwise $d_{K|H}(k) = 0$. Similarly, $D_{M|H}(m) = 0$ if $m < 0$, $m\,(m+1)/2$ if $0 \le m \le 1$ and 1 if $m > 1$; $d_{M|H}(m) = m + 1/2$ if $0 \le m \le 1$, otherwise zero. Next, from (7.63), $d_{K|M,H}(k,m) = d_{K,M|H}(k,m)/d_{M|H}(m) = (k+m)/(m+1/2)$ if $0 \le k \le 1$, otherwise zero. At $m = 0.5$, $d_{K|(M=0.5)\wedge H}(k) = k + 1/2$ if $0 \le k \le 1$, otherwise zero.

2. Extending the argument as in (7.45), the integral is

$$d_{A,C|H}^1(a,c) = d_{A|H}^1(a)\, d_{C|A,H}^1(c,a)$$

$$= \left[\{\Gamma(10)\}^{-1} a^9 e^{-a}\right]\left[\{\Gamma 100\}^{-1} a^{100} c^{99} e^{-ac}\right].$$

$$\therefore d_{C|H}^1(c) = \{\Gamma(10)\,\Gamma(100)\}^{-1} c^{99} \int_0^\infty a^{109} e^{-a(1+c)}\, dx\,.$$

To integrate, either apply the change of variable $x = a\,(1+c)$, or use the fact that the $Ga(1+c, 110)$ density must integrate to one. We find that $d_{C|H}^1(c) = \{B(10, 100)\}^{-1} c^{99} (1+c)^{-110}$ for $c \ge 0$, using also (7.27). Now apply Bayes' theorem, (7.46), to give

$$d_{A|C,H}^1(a,c) = d_{A|H}^1(a)\, d_{C|A,H}^1/d_{C|H}^1(c)$$

$$= \{\Gamma(110)\}^{-1} (1+c)^{110} a^{109} e^{-a(1+c)}\,,$$

i.e. $A|C,H \underset{1}{\sim} Ga(1+c, 110)$. $\therefore A|H \wedge (C=17) \underset{1}{\sim} Ga(18, 110)$.

3.

$$E(X+Y|H) = \int\limits_{-\infty}^{\infty}\int\limits_{-\infty}^{\infty} (x+y)\,d_{X,Y|H}(x,y)\,dx\,dy$$

$$= \int\limits_{-\infty}^{\infty}\int\limits_{-\infty}^{\infty} x\,d_{X,Y|H}(x,y)\,dx\,dy + \int\limits_{-\infty}^{\infty}\int\limits_{-\infty}^{\infty} y\,d_{X,Y|H}(x,y)\,dx\,dy .$$

Take the first term,

$$\int\limits_{-\infty}^{\infty}\int\limits_{-\infty}^{\infty} x\,d_{X,Y|H}(x,y)\,dx\,dy = \int\limits_{-\infty}^{\infty} x\,d_{X|H}(x)\left[\int\limits_{-\infty}^{\infty} d_{Y|X,H}(y,x)\,dy\right]dx$$

$$= \int\limits_{-\infty}^{\infty} x\,d_{X|H}(x)\,dx \qquad\qquad (A.2)$$

since the integral from $-\infty$ to ∞ of any density is one. This is $E(X|H)$ and the second term is similarly $E(Y|H)$, thus proving (6.8). Then if X and Y are independent given H,

$$E(XY|H) = \int\limits_{-\infty}^{\infty}\int\limits_{-\infty}^{\infty} xy\,d_{X|H}(x)\,d_{Y|H}(y)\,dx\,dy$$

$$= \int\limits_{-\infty}^{\infty} x\,d_{X|H}(x)\left[\int\limits_{-\infty}^{\infty} y\,d_{Y|H}(y)\,dy\right]dx$$

$$= \int\limits_{-\infty}^{\infty} x\,d_{X|H}(x)\,E(Y|H)\,dx = E(Y|H)\int\limits_{-\infty}^{\infty} x\,d_{X|H}(x)\,dx$$

and (6.10) is proved. The final part is proved by the same argument as (A.2).

7(e)

1. From Exercise 7(c)8, $S|H \wedge (N=n)\ _1{\sim}$ Gq(0.2n, 1).

$$\therefore d^1_{S,(N=n)|H}(s) = P_1(N=n|H)\,d^1_{S|H \wedge (N=n)}(s)$$

$$= (1/3)\,\{0.2\,n\,\exp(-0.2\,ns)\} = n\,\exp(-0.2\,ns)/15$$

for $s \geq 0$, $n = 1, 2, 3$. From (7.63) and (7.64), $d^1_{S|H}(s) = \Sigma^3_{n=1} n\,\exp(-0.2\,ns)/15$, $\therefore P_1(N=n|(S=s)\wedge H) = \{n\,\exp(-0.2\,ns)/15\}/\{\Sigma^3_{m=1} m\,\exp(-0.2\,ms)/15\}$. Given $S=5$, the denominator is $\Sigma^3_{m=1} m\,\exp(-m) = e^{-1} + 2e^{-2} + 3e^{-3} = 0.7879$ and therefore $P_1(N=1|(S=5)\wedge H) = e^{-1}/0.7879 = 0.467$. Similarly, $P_1(N=2|(S=5)\wedge H) = 0.343$, $P_1(N=3|(S=5)\wedge H) = 0.190$.

2. Use (7.59). We have $d_{X|H}(x) = \{B(a,b)\}^{-1} x^{a-1}(1-x)^{b-1}$ (for $0 \leq x \leq 1$) and $P(Y=y|H \wedge (X=x)) = x\,(1-x)^y$ (for $y = 0, 1, 2, \ldots$). Therefore $d_{X,(Y=y)|H}(x) = \{B(a,b)\}^{-1} x^a\,(1-x)^{b+y-1}$. The marginal distribution of Y is

$$P(Y=y|H) = \int\limits_0^1 d_{X,(Y=y)|H}(x)\,dx$$

$$= \frac{B(a+1,b+y)}{B(a,b)} = a\,\frac{\Gamma(b+y)\,\Gamma(a+b)}{\Gamma(b)\,\Gamma(a+b+y+1)}\ .$$

Using (7.65) we now have $d_{X|(Y=y)\wedge H}(x) = \{B(a+1,b+y)\}^{-1}x^a\,(1-x)^{b+y-1}$.

7(f)

1. $E(X|H) = 625 \times 0.1 = 62.5$, $var(X|H) = 625 \times 0.1 \times 0.9 = 56.25$.
$\therefore sd(X|H) = 7.5$.
From the normal limit theorem, $(X - 62.5)/7.5$ is approximately $N(0,1)$. There-
fore $P(X > 77|H) = P(X > 77.5|H) \approx P(N(0,1) > (77.5 - 62.5)/7.5) = 1 - \Phi(2)$
$= 0.0228$.

2. Use the normal-normal theorem with $m = -36$, $v = 16$, $w = 3^2 = 9$.
Then $T|H \wedge (X = -35)$ $_1 \sim N(m_1, v_1)$, where $v_1 = (16+9)^{-1}\,16 \times 9 = 5.76$,
$m_1 = (16+9)^{-1}\,(16 \times (-35) + 9 \times (-36)) = -35.36$. Standardizing, we obtain
$P_1(T > -25|H \wedge (X = -35)) = P(N(0,1) > (35.36 - 25)/2.4) = 1 - \Phi(4.3)$, which
from Table 7.1 is obviously very small, much less than $1 - \Phi(2.5) = 0.0062$. The
engineer should now be very confident that the liquid will be satisfactory.

3. Using (7.59) the joint semi-density is

$$d_{X,(Y=y)|H}(x) = \left[a^b\,x^{b-1}\,e^{-ax}\,/\Gamma(b)\right]\left[x^y\,e^{-x}\,/y!\right]$$
$$= a^b\,\{y!\,\Gamma(b)\}^{-1}x^{b+y-1}\,e^{-(a+1)x}\ . \tag{A.3}$$

From (7.62), we obtain $P(Y=y|H)$ by integrating this with respect to x. The
change of variable $z = (a+1)x$ yields

$$P(Y=y|H) = a^b\,\{y!\,\Gamma(b)\}^{-1}(a+1)^{-(b+y)}\,\Gamma(b+y) \tag{A.4}$$

and (7.75) is proved. Dividing (A.3) by (A.4) gives the conditional density of X
given $(Y=y)\wedge H$ (from (7.65)) which yields (7.76).

8(a)

1. Whilst $P_1(R_1|H) = 0.75$, you would measure $P(R_2|R_1 \wedge H)$ to be greater
than 0.75, because each rat showing a positive response suggests that the drug is
more effective and hence increases your probability that others will show posi-
tive responses. Similarly, $P(R_3|R_1 \wedge R_2 \wedge H) > P(R_2|R_1 \wedge H)$. Continuing in
this way, your measurement of $P(\wedge_{i=1}^{6} R_i|H)$ will certainly exceed 0.75^6.

2. No. Your probability for the dollar falling on day 2 will be greater if it fell
on day 1 than if it rose.

3.
$P(V_5 \wedge V_6 \wedge ((\neg V_7) \vee (\neg V_8))|H) = P\{(V_5 \wedge V_6 \wedge (\neg V_7)) \vee (V_5 \wedge V_6 \wedge (\neg V_8))|H\} =$
$P(V_5 \wedge V_6 \wedge (\neg V_7)|H) + P(V_5 \wedge V_6 \wedge (\neg V_8)|H) - P(V_5 \wedge V_6 \wedge (\neg V_7) \wedge (\neg V_8)|H)$.
The first term is $P(V_5 \wedge V_6|H) - P(V_5 \wedge V_6 \wedge V_7|H) = q_2 - q_3 = 0.02$. The second
term is also 0.02 because of exchangeability. The third is found to be

$q_2 - 2q_3 + q_4 = 0.01$. The required probability is therefore $0.02 + 0.02 - 0.01 = 0.03$. $P(V_3|V_1 \wedge (\neg V_2) \wedge H) = P(V_1 \wedge (\neg V_2) \wedge V_3|H)/P(V_1 \wedge (\neg V_2)|H) = (q_3 - q_2)/(q_1 - q_2) = 2/3$.

4. Suppose first that $E = (\wedge_{i=1}^{m} R_i) \wedge (\wedge_{i=m+1}^{n} R_i)$. Then applying the product theorem we find

$$P(E|H) = \frac{1}{1} \times \frac{2}{3} \times \cdots \times \frac{m}{m+1} \times \frac{1}{m+2} \times \frac{2}{m+3} \times \cdots \times \frac{n-m}{n+1}$$

$$= \frac{m!\,(n-m)!}{(n+1)!} = \frac{1}{(n+1)\binom{n}{m}} .$$

Now let E be any other proposition asserting that a specified m R_is are true and the remainder false. Applying the product theorem as before, we find first that the denominators form $(n+1)!$ again. The numerators form $m!\,(n-m)!$ again also, but in a different sequence. This is because when the j-th red/white ball is drawn there must be j red/white balls in the bag.

5. $q_1 = P(R_1|H) = 1/2$, $q_2 = P(R_1 \wedge R_2|H) = (1/2) \times (2/3) = 1/3$, et cetera. In general, $q_m = 1/m$. Consider $P(M = m|H)$. We found in Exercise 4 that for any proposition E asserting that a specified m R_is are true and the remainder false, $P(E|H) = \{(n+1)\binom{n}{m}\}^{-1}$. Now there are $\binom{n}{m}$ such propositions agreeing with the proposition '$M = m$'. Therefore, by the sum theorem, $P(M = m|H) = \binom{n}{m}\{(n+1)\binom{n}{m}\}^{-1} = (n+1)^{-1}$ for all $m = 0, 1, 2, \ldots, n$.

6. Use (8.16). For instance, $P_d(X = 0|H) = 0.25 \times \binom{0}{0} \times \binom{32}{12}/\binom{32}{21} + 0.4 \times \binom{1}{0} \times \binom{31}{12}/\binom{32}{12} + \ldots + 0.05 \times \binom{4}{0} \times \binom{28}{12}/\binom{32}{12}$. The computations are quite substantial. The results are

x	0	1	2	3	4
$P_d(X = x|H)$	0.6063	0.3118	0.0707	0.0105	0.0007

8(b)

1. We could proceed from scratch as in Exercise 8(a)4, but following the hint we can use the original urn scheme. Let $R_1, R_2, \ldots$ be the draws from the modified urn, and think of these as $R_1 = R_{a+b-1}^{*}$, $R_2 = R_{a+b}^{*}, \ldots$, where the R_i^{*}s are the draws form the original urn. Probabilities for the modified scheme are as for the original urn, conditional on $R_1^{*}, \ldots, R_{a+b-2}^{*}$ yielding $a-1$ red and $b-1$ white balls. Since the R_i^{*}s are exchangeable, the conditional exchangeability theorem shows that the draws in the modified scheme are also exchangeable.

2. From (8.20), $q_2 - q_1^2 = E(T^2|H) - \{E(T|H)\}^2 = var(T|H) \geq 0$. $P(E_i|H \wedge E_j) = P(E_i \wedge E_j|H)/P(E_j|H) = q_2/q_1 \geq q_1^2/q_1 = q_1 = P(E_i|H)$.

3. We have an infinite exchangeable sequence of possible experiments $F_1, F_2, \ldots$ and will observe F_1, F_2 and F_3. Letting T be the proportion of true propositions in the sequence, the physicist knows that $T=1$ or $T=1/2$. E is the proposition '$T=1$'. We apply Bayes' theorem, (8.22). We easily see that if any F_i is observed to be false then the posterior probability of E (or '$T=1$') becomes zero. Otherwise

$$P(T=1|H \wedge F_1 \wedge F_2 \wedge F_3) = \frac{P(T=1|H)}{P(T=1|H)+(1/2)^3 P(T=1/2|H)} .$$

4. Let H denote your original information. $P_1(T_1|H)=P_1(T_2|H)=P_1(T_3|H)=1/3$. Your new information is $H_1 = H \wedge (X=10)$. Using (8.22)

$$P_1(T_1|H_1) = \frac{(1/3) \times 1}{(1/3) \times 1 + (1/3) \times (1/2)^{10} + (1/3) \times 0} = 0.999 .$$

Similarly, $P_1(T_2|H)=0.001$, $P_1(T_3|H)=0$.

8(c)

1. First we select a beta distribution to represent your initial beliefs. Let $T|H_1 \sim Be(p,q)$, where p and q satisfy $E_1(T|H)=p/(p+q)=0.4$ and $var_1(T|H)=pq/\{(p+q)^2(p+q+1)\}=0.02$. $\therefore p+q+1=0.4 \times 0.6/0.02=12$. $\therefore p=0.4 \times 11=4.4$, $q=6.6$. After observing $X=11$ from 20 trees, $T|H \wedge (X=11)_1 \sim Be(p+11,q+9)=Be(15.4, 15.6)$. Summarizing this, your updated distribution is unimodal with mean $15.4/31=0.497$ and variance $0.497 \times 0.503/32=0.0078$, i.e. a standard deviation of 0.088. Thus, for instance, $P_1(0.409 < T < 0.585|H \wedge (X=11)) \approx 0.65$.

2. Your own probabilities are required. For the final part, $P(S_i|H)=\binom{6}{3}(1/2)^3(1/2)^3=0.3125$. Having done the calculation, no further observations will change your probability 0.3125 for each unobserved S_i. Therefore the S_is become independent and T is now known to be exactly 0.3125. The calculation is equivalent to observing the whole sequence. (Did your initial beliefs about T give a reasonable prior probability to values of T around 0.3125?)

3. $Be(1,1) \sim Uc(0,1)$, a uniform distribution. In Exercise 8(a)5 we showed that your beliefs about M in a finite sequence of Polya urn draws are that it has the uniform distribution on $0, 1, \ldots, n$. Letting $n \to \infty$ produces the $Uc(0,1)$ distribution for T. Alternatively, note that (8.29) with $a=b=1$ agrees with the conditional probabilities for Polya's urn. For general a and b, (8.29) agrees with conditional probabilities for the modified Polya's urn of Exercise 8(b)1. Therefore draws from this modified urn are equivalent to a $Be(1,b)$ initial distribution for T.

4. Fitting a beta distribution to $E_1(A|H) = 0.9$ and $sd_1(A|H) = 0.05$ we find $A|H_1 \sim Be(31.5, 4.5)$. $\therefore A|H \wedge (X=198)_1 \sim Be(229.5, 21.5)$. From (8.30),

$$P(Y=6|H \wedge (X=198)) = \binom{6}{6} \frac{B(229.5+6, 21.5)}{B(229.5, 21.5)}$$

$$= \frac{\Gamma(235.5)}{\Gamma(229.5)} \times \frac{\Gamma(251)}{\Gamma(257)} = 0.5876.$$

9(a)

1. No. The R_is are equi-probable given $H \wedge (S=s) \wedge (V=v) \wedge (C=c)$, but the joint distribution of R_i and R_j given H depends on $|i-j|$. In particular, their correlation coefficient is $c^{|i-j|}$. Exchangeability demands that joint distributions of all pairs should be the same.

2. (a) Yes, unless you had relevant information about the environments in which individual plants grew. (b) Not if you know the sexes of any member of Parliament, because you would not regard women members' heights as exchangeable with men members' heights. (c) Not if you know the locations where individual measurements were made, because you would expect a pair of neighbouring measurements to be more similar than two well-separated measurements. (This is similar to (9.5).)

9(b)

1. The empirical distribution function $f(x)$ is zero for $x < 16.8$, which is the smallest observation. It steps up to $1/30$ at $x=16.8$ and to $1/15$ at $x=17.1$. There are two observations equal to 17.3, so $f(17.3) = 4/30 = 2/15$. And so on. The final step from $29/30$ to 1 occurs at $x = 20.9$, the largest observation. After ten observations the sample mean is $\bar{x}_{10} = (17.3 + 18.4 + ... + 20.5)/10 = 18.97$, and the sample variance is $s_{10}^2 = 1.9941$. Using the whole sample we find $\bar{x}_{30} = 18.793$, $s_{30}^2 = 1.199$.

2. Given $H_1 = H \wedge (T(.)=t(.))$, the X_is are independent. Therefore, $cov(X_i, X_j|H_1) = 0$. $\therefore E\{cov(X_i, X_j|H \wedge T(.)|H\} = 0$. From (9.14), $E(X_i|H_1) = E(X_j|H_1) = E(X_i|H \wedge (M_X=m)) = m$, therefore $cov\{E(X_i|H \wedge T(.)), E(X_j|H \wedge T(.))|H\} = cov(M_X, M_X|H) = var(M_X|H)$. Adding these two parts, $cov(X_i, X_j|H) = 0 + var(M_X|H)$.

3. (9.18) shows that $cov(X, Y|H) \geq 0$. $\therefore corr(X, Y|H) \geq 0$. (In fact, using also (9.17), $corr(X, Y|H) = \{1 + E(V_X|H)/var(M_X|H)\}^{-1}$.) The condition does not hold for a finite sequence. With $n=2$ random variables only, their joint distribution can be anything at all, and can certainly have negative correlation.

4. Your own measurements are required.

9(c)

1. Applying (9.30) we have $m_0 = 100$, $w_0 = 400$, $v = 100$,
$\bar{x}_n = (77 + 65 + 91 + 85 + 78)/5 = 79.2$. $\therefore w_n = 100 \times 400/2100 = 19.05$,
$m_n = (100 \times 100 + 79.2 \times 2000)/2100 = 80.19$.
His posterior distribution is $L|H_n \sim N(80.19, 19.05)$. Therefore
$P_1(L < 100|H_n) = P(N(0, 1) < (100 - 80.19)/19.05^{1/2}) = \Phi(4.54)$, which is very
close to one. He can claim $L < 100$ with a high degree of certainty.

2. Using the expectation of an expectation theorem, and corresponding results
for variance and covariance,

$$E(X_i|H) = E\{E(X_i|M \wedge H)|H\} = E(M|H) = m_0,$$

$$var(X_i|H) = E\{var(X_i|M \wedge H)|H\} + var\{E(X_i|M \wedge H)|H\}$$

$$= E(v|H) + var(M|H) = v + w_0,$$

$$cov(X_i, X_j|H) = E\{cov(X_i, X_j|M \wedge H)|H\} +$$

$$cov\{E(X_i|M \wedge H), E(X_j|M \wedge H)|H\}$$

$$= E(0|H) + cov(M, M|H) = 0 + var(M|H) = w_0.$$

$\therefore corr(X_i, X_j|H) = w_0/(v + w_0)$. (These are also obtainable from (9.16), (9.17)
and (9.18).)

3. Use (9.42). $M|H_n \sim N(3.2, 0.05)$. A standard deviation of 0.1 means a vari-
ance of 0.01. From (9.42) with $v = 1$, you require a total sample size of $n = 100$,
i.e. 80 more observations.

9(d)

1. Let $A|H \sim Ga(a, b)$ and solve $b/a = 1.5$, $b/a^2 = 2$. $\therefore a = 1.5/2 = 0.75$,
$b = 1.5 \times 0.75 = 1.125$. Using (9.45) we note that $n\bar{x}_n = 0 + 1 + 0 + ... + 1 + 1 = 16$.
$\therefore A|H_n \sim Ga(15.75, 17.125)$. The relative weights of data and prior are $n = 15$
and $a = 0.75$ respectively, i.e. the data have 20 times as much weight as the prior
information. This suggests that the prior is relatively weak. Using the $Ga(0, 0)$
prior distribution gives a posterior distribution $Ga(15, 16)$. The two posterior
distributions are very similar. Their means are $17.125/15.75 = 1.087$ and
$16/15 = 1.067$, and their variances are 0.069 and 0.071.

2. $E(\bar{X}_n M_X|H) = E\{E(\bar{X}_n M_X|H \wedge M_X)|H\} = E(M_X^2|H)$. Since $E(\bar{X}_n|H) =$
$E\{E(\bar{X}_n|H \wedge M_X)|H\} = E(M_X|H)$, we have
$cov(\bar{X}_n, M_X|H) = E(M_X^2|H) - \{E(M_X|H)\}^2$.

3. Use (9.59) with the values $m = 0.2$, $w = 0.008$, $v = 0.005$. Then
$c = 25 \times 0.008/(0.005 + 25 \times 0.008) = 0.9756$. The linear estimator gives the estimate $0.0244 \times 0.2 + 0.9756 \times 0.34 = 0.337$.

Index

Pages numbered in bold include definitions or theorems.